T0212333

Research Ethics in Human Geography

This book explores common ethical issues faced by human geographers in their research. It offers practical guidance for research planning and design that incorporates geographic disciplinary knowledge to conceptualise research ethics.

The volume brings together international insights from researchers in geography and related fields to provide a comprehensive overview of relevant ethical frameworks and challenges in human geography research. It includes in-depth reflections on a range of ethical dilemmas that arise in certain contextual conditions and spatial constructions that face those researching and teaching on spatial dimensions of social life. With a focus on the increased need for specialist ethics training as part of postgraduate education in the Humanities and Social Sciences and the necessity for fostering sensitivity in cross-cultural comparative research, the book seeks to enable people to engage in ethical decision-making and moral reasoning while conducting research. Chapters examine the implications of geographical research for conceptualising ethics and discuss specific case studies from which more general conclusions, linked to conceptual debates, are drawn.

As a research-based reference guide for tackling ethically sensitive projects and international differences in legal and institutional standards and requirements, the book is useful for postgraduate and undergraduate students as well as academics teaching at senior levels.

Sebastian Henn holds the Chair in Economic Geography at Friedrich Schiller University in Jena. His research interests focus on knowledge transfers over geographical distance, urban economies as well as on migration and regional development.

Judith Miggelbrink holds the Chair in Human Geography at Technische Universität Dresden. Her research focuses on social geography and globalisation. Currently, her projects deal with securitisation in border regions, cross-border medical practices, peripheralisation and regionalisation. Her methodological focus is on qualitative methods as well as mixed methods.

Kathrin Hörschelmann is Professor of Cultural Geography at the University of Bonn, Germany. Her research focuses on the entangled geographies of (in)security, with a particular interest in childhood and youth. She is co-author of "Children, Youth and the City" and co-editor of "Spaces of Masculinities."

Routledge Studies in Human Geography

This series provides a forum for innovative, vibrant, and critical debate within Human Geography. Titles will reflect the wealth of research which is taking place in this diverse and ever-expanding field. Contributions will be drawn from the main sub-disciplines and from innovative areas of work which have no particular sub-disciplinary allegiances.

British Migration
Globalisation, Transnational Identities and Multiculturalism
Edited by Pauline Leonard and Katie Walsh

Why Guattari? A Liberation of Cartographies, Ecologies and Politics
Edited by Thomas Jellis, Joe Gerlach, and John-David Dewsbury

Object-Oriented Cartography
Maps as Things
Tania Rossetto

Human Geography and Professional Mobility
International Experiences, Critical Reflections, Practical Insights
Weronika A. Kusek and Nicholas Wise

Geographies of the Internet
Edited by Barney Warf

Locating Value
Theory, Application and Critique
Edited by Samantha Saville and Gareth Hoskins

Spatialized Islamophobia
Kawtar Najib

For more information about this series, please visit: www.routledge.com/Routledge-Studies-in-Human-Geography/book-series/SE0514

Research Ethics in Human Geography

Edited by
Sebastian Henn, Judith Miggelbrink and
Kathrin Hörschelmann

Routledge
Taylor & Francis Group

LONDON AND NEW YORK

First published 2022
by Routledge
2 Park Square, Milton Park, Abingdon, Oxon OX14 4RN

and by Routledge
605 Third Avenue, New York, NY 10158

Routledge is an imprint of the Taylor & Francis Group, an informa business

British Library Cataloguing-in-Publication Data
A catalogue record for this book is available from the British Library

Library of Congress Cataloging-in-Publication Data
A catalog record has been requested for this book

ISBN: 978-1-138-58041-1 (hbk)
ISBN: 978-1-032-11679-2 (pbk)
ISBN: 978-0-429-50736-6 (ebk)

DOI: 10.4324/9780429507366

Typeset in Times NR MT Pro
by KnowledgeWorks Global Ltd.

Contents

List of figures		vii
List of tables		viii
List of boxes		ix
List of contributors		x

1 Reflecting research ethics in human geography: A constant need   1
JUDITH MIGGELBRINK, KATHRIN HÖRSCHELMANN AND SEBASTIAN HENN

PART I
Ethics in human geographical research   21

2 Caring about research ethics and integrity in human geography   23
IAIN HAY AND MARK ISRAEL

3 Research ethics in human and physical geography: Ethical literacy,
the ethics of intervention, and the limits of self-regulation   42
SUSANN SCHÄFER

4 Childhood is a foreign country? Ethics in socio-spatial
childhood research as a question of 'how' *and* 'what'   59
KATHRIN HÖRSCHELMANN

5 Ethical challenges arising from the vulnerability of refugees
and asylum seekers within the research process   78
DORIT HAPP

6 Research ethics and inequalities of knowledge production
in Eastern Europe and Eurasia   92
KRISTINE BEURSKENS, MADLEN PILZ AND LELA REKHVIASHVILI

7 Sensitive topics in human geography: Insights from research
on cigarette smugglers and diamond dealers 114
BETTINA BRUNS AND SEBASTIAN HENN

8 Volunteer-practitioner research, relationships and
friendship-liness: Re-enacting geographies of care 133
MATEJ BLAZEK AND KYE ASKINS

PART II
Research ethics in the wider academic context 151

9 Illegal ethnographies: Research ethics beyond the law 153
THOMAS DEKEYSER AND BRADLEY GARRETT

10 Researcher trauma: Considering the ethics, impacts and
outcomes of research on researchers 168
DANIELLE DROZDZEWSKI AND DALE DOMINEY-HOWES

11 Practical ethics approaches for engaging ethical issues
in research geography 182
FRANCIS HARVEY

12 Facing moral dilemmas as a method: Teaching ethical
research principles to geography students in higher education 200
JEANNINE WINTZER AND CHRISTOPH BAUMANN

13 Doing geography in classrooms: The ethical dimension
of teaching and learning 213
MIRKA DICKEL AND FABIAN PETTIG

14 Ethics of reflection: A directional perspective 223
MATTHEW G. HANNAH

Index 240

Figures

12.1 Graduated model of development.   205

Tables

3.1   Differences between self-regulation and external regulation   53

Boxes

2.1 Examples of difficult ethical situations encountered by
     human geographers   25
12.1 Construction principles of a moral dilemma   208
12.2 Preparation and evaluation of soil erosion and land use maps   208
12.3 Sustainable agriculture and food security   209
12.4 Consequences of asylum status for refugee families   209

Contributors

Kye Askins is committed to social and environmental justice, campaigns/is activist on a range of issues, and aims to live as lightly on the earth as she can. She left academia in September 2019 to learn and live on organic and permaculture farms, and with/in alternative communities.

Christoph Baumann did his doctoral thesis at the University of Erlangen-Nuremberg about idyllic ruralities. He works both as university lecturer and high school teacher of geography and philosophy/ethics. His research interests lie in the field of social/cultural geographies, media geographies and geographical education.

Kristine Beurskens coordinates the research group "Geographies of belonging and difference" at the Leibniz Institute for Regional Geography in Leipzig. In her social and political geographical research, she focuses on bordering processes and insecurities, emotional geographies and populism, as well as current debates on qualitative research methods, research practice and visualisations.

Matej Blazek is a social and political geographer with interests in childhood, emotions, marginalised migrant lives and Eastern Central Europe.

Bettina Bruns is a Schumpeter Fellow at the Leibniz Institute for Regional Geography in Leipzig. Her research interests include borders, security studies and ethnographic methods. She holds a Diploma in Cultural Studies from the European University Viadrina Frankfurt (Oder) and a PhD from the Faculty of Sociology of the University of Bielefeld.

Thomas Dekeyser is a British Academy Postdoctoral Fellow at the Centre for the GeoHumanities at Royal Holloway, University of London. His research interrogates urban subcultures, technology, refusal, and negativity, and he is currently writing a philosophical history of resistance to technology.

Mirka Dickel holds a professorship for Geography Education at the Institute of Geography at the University of Jena. Her scientific main focus is on the intersection of philosophical hermeneutics, cultural, social and visual spatialities.

Dale Dominey-Howes is a geographer based at the School of Geosciences at The University of Sydney. His interests and expertise are in environment and society in relation to hazards, disasters and risk. He has published extensively on environmental and human dimensions of hazard and disaster.

Danielle Drozdzewski is an Associate Professor in Human Geography at Stockholm University. She researches across cultural geography, cultural memory and geographies of identities to explore how a politics of memory and identity operationalises spatially. She is Editor-in-Chief of Emotion, Space & Society.

Bradley Garrett is a geographer based at University College Dublin. He has published over 50 journal articles and book chapters. His most recent book is "Bunker: Building for the End Times," which was published in August 2020 by Penguin (UK/Commonwealth) and Simon & Schuster (USA).

Matthew Hannah holds the Chair in Cultural Geography at the University of Bayreuth. His research focuses on spatial power relations, state knowledge and critical social theory.

Dorit Happ is an independent migration researcher. She received her doctorate in geography from the Friedrich Schiller University Jena in 2020. Her current research project assesses the labour market integration of refugees and asylum seekers in Berlin.

Francis Harvey is lead of the data and practices group at the Leibniz Institute of Regional Geography in Leipzig. His current research interests focus on spatial/temporal representation of historical phenomena. He is author of the book "A Primer of GIS", published by Guilford Press.

Iain Hay is First Vice-President of the International Geographical Union and Matthew Flinders Distinguished Emeritus Professor of Geography at Flinders University. He has published extensively on geographies of oppression and domination, critical economic and cultural geographies, professional ethics and geographical education.

Sebastian Henn holds the Chair in Economic Geography at Friedrich Schiller University in Jena. His research interests focus on knowledge transfers over geographical distance, urban economies as well as on migration and regional development.

Kathrin Hörschelmann is Professor of Cultural Geography at the University of Bonn. Her research focuses on the entangled geographies of (in) security, with a particular interest in childhood and youth. She is co-author of "Children, Youth and the City" and co-editor of "Spaces of Masculinities."

Mark Israel is a senior consultant with Australasian Human Research Ethics Consultancy Services and Adjunct Professor at the University of

Western Australia. He has qualifications in law, sociology, criminology and education and has published on research ethics and integrity, higher education practice and criminology.

Judith Miggelbrink holds the Chair in Human Geography at Technische Universität Dresden. Her research focuses on social geography and globalisation. Currently, her projects deal with securitisation in border regions, cross-border medical practices, peripheralisation and regionalisation. Her methodological focus is on qualitative methods as well as mixed methods.

Fabian Pettig is an assistant professor at the University of Graz. His research focuses on critical and emancipatory geography education with a dedicated interest in the geographical and pedagogical implications of sustainability and digitality.

Madlen Pilz is currently a researcher at the Leibniz Institute for Research on Society and Space in Erkner/Berlin. Her main research topics are critical urban and migration studies, post-socialist spaces, everyday life studies and protest practices.

Lela Rekhviashvili is a researcher at Multiple Geographies of Regional and Local Development research area at the Leibniz Institute for Regional Geography, an invited lecturer at Universities of Leipzig and Jena. She is a political economist, researching urban informality, transport and mobility, marketisation and its contestations.

Susann Schäfer is a postdoctoral researcher in economic geography at Friedrich Schiller University Jena. Her research interests focus on entrepreneurial ecosystems, climate change adaptation, and new methods in human geography.

Jeannine Wintzer has been a lecturer in qualitative methods at the Institute of Geography at the University of Berne since 2013. Her research topics are visual geography, discourse, narrative, argumentation and metaphor analysis, critical population and geographical science research. In addition to qualitative methods in geography, she teaches history of geography.

1 Reflecting research ethics in human geography

A constant need

Judith Miggelbrink, Kathrin Hörschelmann and Sebastian Henn

Introduction

Human geographers deploy a broad variety of conceptual and methodo-logical approaches and use different methods to understand the interrela-tionships between people, place and space. As there is no uniform agenda or shared conceptual position, research in human geography has to tackle a number of ethical challenges that result from the various ways in which research affects individuals and social groups, their physical, mental and economic well-being, their reputation and social positions as well as their image in a wider societal context. Human geographers are, of course, not the only ones facing such challenges. Rather, doing research in an ethi-cally acceptable way has become a significant concern for empirical disci-plines in general and for those involved in related research activities. This includes researchers aware of their practices' effects on the contexts and subjects of their research, legislative bodies identifying, defining and sanc-tioning illegal practices, communities counteracting exploitive, colonialist, racist, ethnocentric, androcentric and other non-sensitive research prac-tices they experienced in the past, and third-party funders influencing and controlling the way their money is spent.

In this introductory chapter, we provide an overview of key ethical issues that we consider particularly relevant for research in human geography today. We start with the assumption that human geographers are increas-ingly confronted with ethical issues, particularly the principle of not caus-ing harm and the necessity to obtain informed consent. Proceeding from an overview of debates on ethical challenges in human geography research that are embedded in broader and transdisciplinary strands of reasoning, this introduction pays special attention to five major topics: questions of power and social relationships in research processes; ethical challenges arising from a wider neoliberal research context in which many of us find themselves positioned; ethical requirements of handling research data; institutional requirements regarding the implementation of research eth-ics; and strengthening ethical reasoning through scientific education. Without claiming to treat the subject conclusively, we ask to what extent it

DOI: 10.4324/9780429507366-1

is necessary to consider research ethics through the particular lens of our discipline. We conclude that translations and adjustments from other fields of knowledge are necessary in order to tackle the specificities of research in human geography.

Basic principles in research ethics

Reflections on ethical questions that arise in the process of planning and conducting research have accordingly become a key concern in scientific discourse (e.g. Sultana 2007; Zhang 2017). Although this discourse has significantly broadened, it is nevertheless striking that two principles that overlap and inform each other typically serve as the entry points for researchers confronted with ethical challenges: The first principle, which could be paraphrased as "do not harm," primarily appeals to the researchers' sense of moral duty. It demands self-commitment and places a high responsibility on individual researchers, even where guided and supported by shared standards that apply to all members of a scientific community. This principle provides a compass to researchers in that it guides their decisions even in the absence of formal institutional rules or in cases where informed consent cannot be obtained (e.g. research about illegal practices).

The second principle, informed consent, is a cornerstone of many in existing – though perhaps disputed – regulatory frameworks that researchers have to comply with in order to fulfil formal research requirements. Such institutionalised ethics help to channel and structure researchers' efforts to act ethically though it may also conflict with their convictions. The differences between these two starting points have been mirrored by broader debates on research ethics. These often view the principle of "not doing harm" as a part of practical ethics, while the principle of "informed consent" is typically regarded as part of a procedural approach that limits researchers' and institutions' liabilities.

A particularly thorny issue that goes directly to the heart of the debate on research ethics is the moral framework that guides researchers' decisions. General guidelines may seem desirable in this context but are more difficult to ascertain than they perhaps first appear. Although it is generally accepted that research practices should follow the principle of "not doing harm," it remains challenging to define the basis of this principle, namely *harmlessness* – both in different situations and for different subjects. It is also debatable whether harmlessness can actually be achieved (cf. Boyd et al. 2008). All research efforts and researchers struggling with ethical issues are confronted with this basic advice – often without knowing exactly at which point harmlessness turns into its opposite. Further, although the principle of not causing harm is emphatically agreed upon by many scholars, it remains "most contested," as Bell (2014: 511) states. Research ethics needs to provide orientation and equip researchers with a toolkit that enables them to reflect on their practices beyond purely legal concerns.

Various debates, shaped by disciplinary differences and other aspects, have shown that it is very difficult to find an understanding of research ethics that is universally accepted. Usually, as Guillemin and Gillam (2004: 262) state, the "starting point and focus for discussions of research ethics has been the ethical principles formulated for biomedical research, usually quantitative in nature." However, this perspective has been criticised as not sufficiently reflecting the specifics of social science research, including human geography (cf. Dyer and Demeritt 2009). It is, therefore, no coincidence that different social science disciplines have developed their own approaches to research ethics. Debates in this regard often oscillate between the search for common ground and specific challenges arising from historically developed specific research practices of the discipline. This volume is no exception to his rule.

Differences in definition and perspective also complicate the application of the *principle of informed consent*. At a basic level, informed consent refers to building and displaying a research relation between researcher and participant that is characterised by the researcher's respect for participants' integrity and awareness of their bodily, mental and social vulnerabilities (cf. Burgess 2007: 2293). Consent, according to this understanding, refers to the state of agreeing on both sides to enter into a social relationship that is based on people accepting each other on equal terms. However, due to histories of exploitation and abuse in the name of science and since research often establishes an unequal if not hierarchical and potentially paternalist relation between researchers and researched, informed consent has become a major area of debate. The "bureaucratic conception," as Burgess (ibid.: 2285) put it, has an "important institutional role, yet may be limited in the extent to which the rules of informed consent promote the substantive conception of relations between researchers and participant." Informed consent is increasingly understood as a documented agreement signed by both parties at the very beginning of fieldwork. Although this formalised pattern of establishing consent may provoke complications that would not have occurred without formalised procedure, it has become a powerful tool for establishing ethical standards in research. As a consequence, research ethics has increasingly come to be understood as an *institutionally designed process* of informing a (potential) research participant about a certain research context, the uses of data and information and how research may affect him/her and/or associated communities. At many universities and funding agencies, this process has become a mandatory precondition for carrying out research and/or the approval of funding.

Research ethics that has become institutionalised in such a way clearly helps to protect participants in research projects. It also offers guidance to researchers and forces them to reflect ethical aspects of their work. In doing so, procedural ethics, however, at the same time shows its Janus-faced nature: On the one hand, it offers researchers formally approved lists of questions that might help them to cut through the thicket of ethical challenges. On the other hand, by facing such streamlined processes of reflection, researchers run the risk of missing the concrete and contextual ethical challenges of a certain project. Despite these shortcomings, there clearly is an increasing relevance of reflecting research ethics in an institutionalised manner, as is, for example, reflected by the formal procedures adopted by the National Research Council of the United States and the Australian National Statement on Research Ethics, or the ethical guidance of UK Research and Innovation (UKRI) for researchers and ethics committees. These developments towards a greater formalisation of research ethics and its institutional governance affect every discipline and thus also human geography.

Research ethics as a part of the research process

Research ethics clearly encompasses more than just the two principles discussed above. In fact, far from being a static set of rules that must be observed while preparing and carrying out research projects it denotes a contested field of laws, interests and orientations that defies simple definitions. Things get even more complicated as the capacity for research to impact both society as a whole and individual lives has expanded remarkably due to technical, economic, social and political transformations. When we think of research as a set of practices that affect people and their living conditions, we need to ask ourselves how far reflections on ethical concerns must reach. Are we prepared and able to fully anticipate the potential impacts of our actions? And how can reflections on research ethics be distinguished from broader debates on moral and ethical concerns? Following this line of reasoning, the contributors to this collection adopt diverse positions with respect to the specific issues they have addressed in their work, and which result from different philosophical perspectives. What unites the different chapters nonetheless is their recognition of the need to continuously reflect on and discuss research ethics, not least because of the changing social conditions and technological shifts that confront us constantly with new challenges.

From our point of view, research ethics is an inherent part of all the different stages of a research process. As such, it has to develop ethical schemes that guide research practices continuously, by reflecting their moral underpinnings. Furthermore, research practices must not be considered an isolated part of academic work for which ethical reflection is most necessary and relevant. Rather, it is intrinsically connected to moral philosophy and needs to inform the latter.

We view it as a necessity to uncover and reflect on the moral underpinnings of our research and to make them transparent in the name of the integrity of the subjects, groups and communities addressed in our work. As we understand it, research ethics is a field of reasoning that spans from the acknowledgement of fundamental inviolable rights (i.e. human, animal and other rights) to the necessity of keeping pace with societal dynamics and contexts to which it needs to be constantly adapted. Research ethics, hence, is not only informed by institutional ethics but also expected to be developed according to the latter, i.e. it must follow institutionalised requirements backed by legal regulations. At the same time, institutional contexts have to accommodate changing ethical concerns. Research ethics is thus institutionally defined in that it sets the limit of what is legal and acceptable. Institutionalisation, however, not only involves what is accepted but also adequate ways to implement and enforce it. Although institutionalised research ethics provides a substantial and powerful regulatory framework, it does not exhaust itself in the same; moreover, not all researchers fully agree with the institutionalisation of ethics. On the contrary, there is substantial criticism of implementing ethical standards in a bureaucratic, administrative manner (Heimer and Petty 2010), and alternative ways of thinking are offered (Dickson and Holland 2017; Dekeyser and Garret in this volume). But what does it mean to act 'alternatively,' when research addresses, and thus itself perhaps borders on, illegal practices? Acting ethically in research in our view does not mean to tick one box or the other. Rather, following Dekeyser and Garrett (this volume) research ethics has to constantly manoeuvre between different understandings of what is considered ethical. This includes addressing the question of whether research into the illegal itself is illegal, whether such research can nonetheless be (seen as) ethical, as well as answering the question of how and when researchers cross a line drawn by the law.

Finally, although there may be consensus on some fundamental aspects, we consider research ethics as contextual for two key reasons. First, if we accept that research ethics is institutionalised (even if we disagree with the way it is done), regulations are typically closely tied to the legal framework of the country in question. However, what is accepted in one country in terms of an institutional approach need not be accepted in another and this is a particularly problematic issue for research conducted across national borders or in the context of international collaborations. Second, debates about research ethics are often partisan and parochial in the sense that they emerge from and relate to a discipline's specific moral economy of attention. Though some ethical standards (e.g. considering needs and demands of a community before defining a research question) can resemble one another, they may differ remarkably in terms of their significance regarding concrete disciplinary and subdisciplinary practices. As the chapters in this volume show, differences in terms of methodological and epistemological choices as well as incorporated patterns of attention and sensitivity mean that human geographers address particular questions and develop specific disciplinary insights (also see Hay and Israel, this volume). In the following sections, we seek to map some of them, while underlining the need for transdisciplinary exchange. More specifically, we ask why we should think about ethical issues *in human geography* and why we should do so (again) now.

Disciplinary debates on research ethics in human geography

Reflecting on research ethics in human geography is not new. About 20 years ago, following David Smith's (1997) landmark review that advocated for a closer engagement of geography with moral philosophy, James Proctor and David Smith published a widely received edited volume on *Geography and Ethics* (1999), which could already build on a remarkably broad literature exploring the "confluence of geography and ethics" (Proctor 1999: 5). This volume clearly marked a milestone in that it brought together numerous essays that addressed ethical implications of geography as a discipline, but also explored ethical issues arising from the discipline's major themes and methodological approaches. Amongst other things, it dealt with inequalities inherent in research relations established through interviews and empirical research in general (Gormley and Bondi 1999; Herman and Mattingly 1999; Rundstrom and Deur 1999; Winchester 1996) as well as with research plans that run counter to participants' interests and demands (also see Wintzer and Baumann, this volume). The volume could build on analyses of ethical issues in research that had been dealt with in handbooks and papers on research methods and professional practice (Eyles and Smith 1988; England 1994; Hay 1998a).

Geographical engagements with ethics and moral philosophy have, since Smith's (1997) call for a 'moral turn,' continued along the dual trajectory of dealing with the substantive moral implications of space on the one hand and of geographical professional practice on the other (cf. Smith 1997, 2000, 2001; Cutchin 2002; Proctor 1999; Barnett 2011, 2012, 2014). Concerning the former, debates on the relationship between morality and space have prompted geographers to increasingly address issues of social justice and spatial representation in terms of the moral geographies underpinning and co-producing them. This has been acknowledged in recent reviews by Olson (2015, 2016, 2018) and is evident from the breadth of work published not least in the journal *Ethics, Place and Environment* (cf. Hay 1998b; Smith 1998; McEwan and Goodman 2010) as well as the special section and a regular feature dedicated to ethics in human geography in the journal *Area*. Research on moral geographies and debates on the relationships between normative reasoning and space has, in the words of Olson (2015: 518), shown that "attentiveness to space has the potential to improve normative reasoning" and to expose "the otherwise hidden spatial dimensions of ethical reasoning" (ibid.; also see Barnett 2016; Olson and Sayer 2009; Popke 2006; McEwan and Goodman 2010). While moral geographies thus reach beyond research ethics, insights from these broader debates are highly relevant for reflecting on the moral reasoning that informs ethical decision-making in research and that geographical professional practice brings to the fore. Thus, geographical research and wider professional practice are informed by moral philosophies of just, unjust, injurious or beneficial practices and they, in turn, inspire discussions on the complexities of such normative reasoning. Participatory action research is perhaps one of the clearest examples of this, as it seeks to address social, spatial and environmental injustices through the conduct of research itself (Cahill et al, 2007; Kindon et al, 2007; Pain 2004; Valentine 2004).

The relevance of broader philosophical debates on the moral complexities of research and wider professional practice is the starting point of the volume that we present here. We understand morality here as the realm of reflecting the normative substance of doings and sayings, and ethics as the area of philosophical reasoning that illuminates how and why moral judgments are applied (e.g. Proctor 1998: 9). Further, we draw on Proctor's helpful threefold distinction between descriptive ethics, normative ethics and meta-ethics: "The aim of descriptive ethics" (ibid.), as Proctor explains, "is to characterize existing moral schemes" (ibid.), whereas "[n]ormative ethics are devoted to constructing a suitable moral basis for informing human conduct" and "[m]eta-ethics, in distinction, is more an examination of the characteristics of ethical reasoning, or systems of ethics" (ibid.). Contributors to our edited volume engage particularly with normative ethics as it pertains to conduct in research. They also reach beyond this specific concern, however, to wider reflections on moral schemes and ethical reasoning.

Focusing specifically on human geography, the chapters collected here reflect in-depth on the difference that space makes to research and to reflections on professional practice in the discipline. They engage closely with ongoing debates in geography on positionality and relations of power in research (Barker and Smith 2001; England 1994; Hopkins 2007; Horton 2008; Kearns 1998; McDowell 1992; Rose 1997; Valentine 2003, 2005; White and Bailey 2004), as well as on the implications of institutional ethics regulations for research practices (also see Dyer and Demerritt 2009; Hay 1998b; Israel and Hay 2006). Our arguments are neither restricted to (human) geography nor do they stem entirely from geographical debates. However, moral and ethical reasoning in geography have evolved in a specific disciplinary context and are affected by the particularities of geographical research. As Proctor (1999: 9) rightly states, "[g]eographical knowledge does not arise in vacuum." Instead, it relates in various ways to the circumstances in which it – literally – takes place. Any disciplinary reflection on research ethics, thus, requires both comprehensiveness and specificity. Given the necessarily selective and non-exhaustive character of an edited volume, we suggest reading the chapters as case studies that, despite their singularity, respond and contribute to key issues that we now outline in turn.

Power and relationships in research processes

Debates on research ethics have to acknowledge that research inherently establishes unequal relations, or, as Richardson and Godfrey (2003: 349) put it with regard to qualitative approaches: "the nature of a relationship between an interviewer and a participant is inherently exploitive. An interview would not happen until the researcher initiated it because they want 'data' – regardless of whether it is characterized in 'exploitive' language as data collection or in amore collaborative language as data generation." Research always comes along with *certain* interests and the initial decision to research a particular topic is most often extrinsic to the people, groups and communities that *become* subjects of research (cf. Beurskens, Pilz and Rekhviashvili, this volume). Hence, with the exception perhaps of participatory and collaborative research, themes, topics, perspectives and methods are more often defined by researchers than by those being 'researched.' Geographic research has, as Rundstrom and Deur (1999: 237) put it, acted across an imaginative, yet real and powerful boundary between the worlds from which western academia has emerged and the worlds that became subject to academic attention and interest – an interest that often manifested itself in an "abusive, colonizing academic gaze and the institutional out of which it peers" (ibid.; also see McEwan and Goodman 2010). Indeed, many authors have pointed to power relations that underly and inform research relations and in this context have looked, for example, towards feminist, indigenous, cultural minority and postcolonial studies for ethical guidance and orientation. As Coombes et al. (2014) have argued with a view to indigenous geographies and the possibilities and limits of community-based participatory research:

> As feminist methodologies shifted the discipline's view of its purpose a generation ago, debates about approaches, collaborations and ethics in Indigenous geographies unsettle our journey as researchers. They reset the compass by which we guide our endeavours and judge our achievements. They invite and challenge, but they also insist on a fundamental reconception of research partnerships and the grounds for mutually beneficial praxis and pedagogies.
>
> (ibid.: 851)

Who speaks in whose name, whom we assume to be a representative speaker of a community and whether we regard a community as a closed entity or as shifting networks of temporarily stabilised relations and interests has become a critical juncture in reflecting *what* research does and *who* performs it. Moreover, as research often subjugates local circumstances under its preformulated epistemological interests, outcomes might not only miss what is relevant for the respective community, but they might also be quite poor. Health research in aboriginal communities is a striking example of this, showing that developing ethical standards and developing *meaningful* research depend on each other (e.g. Flicker and Worthington 2012). The following quote from a contribution on "a regional model for ethical engagement" addresses the problem head-on (Maar et al. 2007: 55):

> Aboriginal communities, government agencies, and academic institutions all require good data sources in order to develop effective interventions; however, conducting research on Aboriginal health issues

has become an immensely complex endeavour that requires special knowledge and training in the areas of Aboriginal health, participatory research methodology, and research ethics. In order to address pressing and immediate health issues properly, researchers must also consider the effects of colonization on Aboriginal community health, including societal power imbalances, the loss of culture and a traditional way of life, and the experience of forced assimilation; many of these health determinants are still poorly understood in the health sciences. The historical circumstances that have shaped Aboriginal health issues also have had a profound impact on acceptable approaches to Aboriginal community research and contemporary Aboriginal research ethics.

Developing (community-based) Aboriginal research ethics is one outcome of long-lasting and fierce struggles for de-colonialisation and self-determination that, amongst other things, include struggles for control over community-related knowledge production (e.g. Brant Castellano 2014). As an outcome of these often painful processes, research has to adjust to changing legal landscapes that concern both community-related rights (e.g. first nations', indigenous and minority rights) and changing individual rights (e.g. for informational self-determination).

While helpful for shifting some of the imbalances that arise from uneven conditions of knowledge production, inherently unequal research relations cannot be entirely countered and healed by participatory approaches and research designs either. At the end of the day, researchers can leave the scene, whereas participants are left with their everyday lives and, moreover, with the potential ramifications of the research encounter. Though this may often not be a problem at all, certain constellations reveal a vulnerability of participants that comes from getting involved in research projects. Projects working with refugees and migration policies thus show clearly how delicate it is to establish research relations in a context that is overwrought by hope and despair, existential fear and exposures to all kinds of humiliation, as well as a lack of resilience (also see Happ, this volume). Researchers bear a lot of responsibility in this regard. However, both sides may be obligated to consider the risks that can emanate from research encounters – including the risk of not being able to estimate what will ensue.

Calculating the potential risks of taking part in a research project is, however, but one aspect influencing the decision of whether or not to grant (informed) consent. Curiosity, pride, expected personal benefits, a desire to help, but also perhaps unwarranted trust, insufficient levels of self-protection and other motives and individual dispositions may also interfere at this point. Bruns and Henn (this volume) thus discuss how participants' interests and their wishes to protect themselves, their business relations, their families and networks are negotiated while establishing trustful relations between researchers and participants. They show that moral assumptions and ethical reasoning are not a distinct step in a line of steps that consecutively follow each other but are constantly negotiated and re-negotiated in an often-time-consuming process that may keep a researcher in a permanent state of uncertainty over whether his or her research plans will work out.

Fieldwork encounters in most cases are not built from scratch but rely on relations that already existed before people are drawn into professional schemes of methodically guided observations. Therefore, it is not enough to establish transparent and fair research relations in the fieldwork encounter when researchers feel involved in more complex relations of engagement and commitment towards more-than-research-subjects. Layers of responsibility force them to constantly negotiate different roles – as a researcher, friend, or professional. Keeping these roles separate when they intersect in the process of building balanced, transparent and respectful research relations can be highly challenging, as Blazek and Askins (this volume) also show.

Research ethics in a wider neoliberal research context

An ethically informed attitude has to consider the rights and positions of *all people involved*. This includes, first and foremost, that participants in a research project "are always treated as an end and never as a means, that is merely an instrument for the achievement of the researcher's ambitions" (Gaudet and Robert 2018: 123). Empirical work, thus, oversteps its ethical limits when following its track regardless of its effects on those who participate. However, empirical research may not only affect those who agreed to take part but also researchers themselves and others who may be involved in the fringes, such as family members or business and community partners. Awareness of potentially negative impacts on researchers themselves has been slowly increasing (cf. Rat für Sozial- und Wirtschaftsdaten 2017: 8) but is still not a central aspect of formal ethics processes. Considering scientific working conditions in terms of ethics might not seem, at first glance, a core aspect of *research* ethics. However, if research is regarded as a socially embedded practice, we cannot ignore the formative effects that conditions of knowledge production have on researchers' subjectivities and their resources for dealing with consequences of research that may be harmful to them or for conducting research that requires particular ethical sensitivity. This means turning "our lens upon our own labour processes, organizational governance and conditions of production" (Gill 2017: 2) in order to identify the risks of being unethical as a structural moment of developing academic careers today. A full-scale examination of processes and conditions of knowledge production and their influence on professional ethics includes reflection on working conditions, on institutional pressures and requirements, on the availability or not of necessary resources and support structures, and even on wider academic practices such as reviews that are, or are not, conducted in ways that respond to the inevitable orientedness of research, as Hannah shows (this volume).

Neoliberal conditions characterise many academic contexts today. They entail precarious working conditions such as limited contracts based on third party funding, a limited number of permanent positions at universities and research institutes and the constant pressure to attract funding and to publish in a highly competitive environment (also see Drozdzewski and Dominey-Howes, this volume), that expose researchers to a double risk. They take on the risk of failing if their research fails. And they take on the risk of not carefully considering the potential harm their work may cause, which in turn may draw into doubt what they had promised in a proposal, a contract or a PhD agreement. As Oberhauser and Caretta (2019: 57) show, mentoring and a critical reflection on the mentoring process can help to identify such risks and to support young academics in particular, by applying "feminist principles of dialogue, reflexivity, and ethics of care as a framework to guide this mentoring process in the challenging landscape of neoliberal higher education." Often informed by feminist ethics of care (Manzi et al. 2019; Horton 2020), researchers increasingly also take the risk of intervening in "the increasing neoliberalisation of academic lives and the intensification of workloads" (Puāwai Collective 2019: 37). Even if one does not subscribe to the diagnosis of the neoliberal university, the discussion it sets in motion is worthy of broader attention as it puts "examples from work and relationships, duties of care, career pathways and strategies, professional and personal development opportunities, travel expectations, and more" (ibid.) on the table in order to challenge and disrupt the widely accepted and taken-for-granted normative settings of today's academy. Ethics of care thus should embrace researchers within their social and institutional contexts as well as those they get in contact with during fieldwork. This includes also a thorough reflection on research topics, especially when young scholars start out on the path of developing academic careers in highly competitive environments. Negotiations of 'relevant' research topics as well as research designs are not infrequently driven by underlying assumptions regarding their impact on one's career or reputation while ethical considerations might be deferred. What might appear as being an individual's decision at first glance, reaches into a broader debate on the normative ethics of *what* should be researched (also see Hörschelmann, this volume). Though a lot of things would be exciting and innovative to investigate, the potential agenda might change if a topic is chosen from an ethically informed point of view. This includes so-called civil clauses, i.e. voluntary commitments by scientific institutions to conduct research exclusively for civil purposes, but also the decision to adjust research to certain community needs or to support positions that are usually marginalised and/or silenced.

However, in a highly competitive environment research ethics might also be invigorated through the backdoor by the imperative of 'publish or perish'. Increasingly, international peer-reviewed journals expect ethical approvals and declarations of no-objection as a pre-condition to start and complete their review processes.[1] Research not meeting these expectations runs the risk of getting excluded from this crucial stage of scientific knowledge production. Ethical expectations articulated in the publishing process might, hence, rebound on the research process itself in terms of being anticipated by researchers who wish to publish their results. Procedural ethics directed to the narrower field of *research* ethics, thus, are not limited to the funding process of a research endeavour but have a formative effect from the stage of developing a research proposal to the dissemination of results.[2]

Research ethics and data handling

Another issue that, at first glance, seems to lie beyond what is commonly thought of as *research* ethics, is the archiving and sharing data (e.g. Irwin 2013). We consider this aspect here as well as however as, for example, the use of stored and archived data relies on the integrity of those who archive, on their abilities to foresee and thus gain informed consent for the future use of research data, and on the ways in which the re-use of archived data is regulated.

Ethical concerns over the storing, archiving and secondary use of data are not new. Researchers dealing with qualitative and archival material have struggled with questions of protecting informants and sources, for example, through anonymising and assuring confidentiality and integrity, for a long time (e.g. Moore 2012). However, there are some recent developments that require closer examination. As funding bodies increasingly require the accessible storage of research data, including transcripts from qualitative research, new questions have arisen that concern not only the archiving process as such but also (potential) 'secondary use.' Two examples identified by Richardson and Godfrey (2003: 348–349) may illustrate our argument here: First, strategies to mitigate the exploitative nature of data collection/generation towards research participant(s) are generally not an option when it comes to secondary use. The use of interview materials may not be as carefully controlled in reuse as researchers had originally assured participant(s) and open-access archiving may undermine ethical standards

---

1   For an overview on ethical issues emerging during publication process see Benos et al. (2005). Though they take their material from physiology, many of their arguments apply to a wider range of sciences.
2   There are of course a lot more ethical challenges luring in the publication process considering the whole process of submitting; allocating to peers and then peer reviewing; communicating between authors, editors and reviewers; and, finally, deciding whether the manuscript will be published or rejected. As geographical journal does not operate audit processes the depend on the reputation and a sense of responsibility of peers who, in turn, have to master the process based on, inter alia, their ethical positions and decisions.

of informed consent. Second, the contextual background may be missed in the secondary appropriation of data. This may lead to misinterpretations that run counter the trust that participants placed in researchers to develop understandings from a position of deeper insight and contextual sensitivity. In other words, any reinterpretation lacks knowledge of the original field research context. Any reuse of the research data could thus prove inadequate in terms of interpretation and the representation of participants' accounts. The more data is generated based on trusting interpersonal relationships, the more ethical issues consequently arise in terms of storage and sharing (also see Hörschelmann, this volume).

Ethics and institutional requirements

As explained in our discussion of the principle of informed consent at the start of this chapter, researchers are widely faced today with requirements by funding bodies and their own academic institutions to reflect on research from an ethical perspective. "Procedural ethics" and "values codified in laws, codes or rules" (Gaudet and Robert 2018: 122), consequently frame and regulate many research ambitions. Research projects in general and, more specifically, fieldwork plans have to be approved by ethics committees – typically located at universities – to ensure that pre-defined standards are met. Often explicated in a catalogue of rules that must be followed, procedural ethics obligate researchers to reflect on and adapt the ethical implications of their actions, even though they are not unproblematic and have been criticised as not meeting the real needs of fieldwork encounters (cf. Happ et al. 2018: 24ff.).

For researchers, formal ethics procedures may sometimes appear as boxes that simply have to be ticked in the right way: Guidelines that should be followed when writing a proposal, questions that need to be answered when applying for funding – a barrier that has to be overcome before the real work starts. However, procedural ethics are neither static nor do they exist in separation from societal, political and economic contexts. Quite the contrary: Procedural ethics imply that there are social entities that develop and set requirements and that have at least some form of control over whether researchers comply with them. Moreover, procedural ethics are bound to broader societal regimes and discourses that define the limits of the acceptable and the legitimate. Deeply embedded in the "current state of society," procedural ethics might be as problematic as the society spawning them. Furthermore, the history of science has shown that procedural ethics tend to have difficulties keeping pace with the times. Rather than anticipating future challenges, they often struggle to adequately deal with innovations and social transformations, as can be seen in the area of genetic engineering or the increasing sensitivity towards the rights of animals and other more-than-humans (Buller 2016; Lynn 1998; Popke 2009; Whatmore 1997). On the other hand, well-established research might be challenged when confronted with ethical requirements that are suddenly applied to the field (cf. Skloot 2010). At the same time, however, new ethical problems might emerge from newly invented fields of research (e.g. stem cell research, big data or applying digital methods).

Research ethics and scientific education

Though moral reasoning often appears as an individual challenge arising from concrete fieldwork practices, researchers can try to prepare themselves and – in as far as they may be academic teachers, mentors and principal investigators – support others to prepare themselves. Such preparation includes specific, project-related considerations as well as wider reflections on the ethical implications of disciplinary practices and histories of knowledge production (also see Wintzer and Baumann, this volume). As geography has been part of a set of disciplines that have accessed places, people and environments in colonial, imperial, and otherwise exploitive ways, university teachers should seek to actively engage students in reflective discussions on the normative implications of our disciplinary history as well as on the ethical issues that arise from fieldwork conducted, for instance, with the aim of gaining a university degree. Exercises enabling students to identify ethically relevant topics and to address them adequately, by stimulating ethical reasoning, can be integrated in various courses, including GIS, as Harvey et al. (this volume) suggest (also see Gannon 2014). Harvey's as well as Wintzer's and Baumann's chapter (this volume) offer practical guidance for this, while Dickel and Pettig (this volume) show how and why ethical reasoning can be implemented in teaching at all levels, including secondary school (also see Kirman 2003).

Human geographical research ethics or research ethics in human geography?

If applied universally, then principles such as the "dignity of human beings, the integrity of people and the values of justice" (Gaudet and Robert 2018: 122) concern all research practices and disciplines. It follows logically that there is no need to adopt separate ethical norms or standards with respect to any particular discipline. Instead of conceptualising a particular 'human geographic' attitude towards research ethics, we regard ethical aspects as indispensable underpinnings of research practices in general and, therefore, also draw on the rich and methodologically specific literature on research ethics in other disciplines (e.g. Gaudet and Robert 2018, Ryen 2010, Silverman 2017, Yin 2018). Furthermore, a transdisciplinary approach to ethical issues is justified by sociotechnical changes, which not only raise new research questions but also drive methodological developments, such as the analysis of data obtained from online media (cf. Madge 2007). Here, too, ethical considerations cannot be confined to the narrow boundaries of a single discipline (cf. Rogers 2019: 208–209, 245–246; Moreno et al. 2013; Fossheim and Ingierd 2015).

However, research practices differ in terms of their disciplinary origins, thematic foci and subject-related conceptual questions. They therefore also vary in terms of how, when, and what kinds of ethical issues arise in research. Certain ethical challenges may be more pressing in some disciplines than in others. When examining research ethics in human geography, characteristics and requirements of research in human geography thus come to the fore more. And, although problems, topics and theories might overlap between different disciplines, debates on ethical issues cannot be transferred seamlessly and in their entirety from other disciplines to human geography. Within geography itself, there are ethical issues that apply more to one subdiscipline than another, although debate across subdisciplinary divides can prompt novel conceptualisations and approaches that oblige us to reconsider our understandings of research ethics too, as current discussions on the more-than-human show (cf. Lynn 1998; Popke 2009; Schäfer, this volume; Whatmore 1997). A 'human ethics' which focuses solely on human beings often falls short of what is required in the field of geography, since both questions concerning the ethical claims of non-human entities (animals, plants, …) and the handling of natural and artificial artefacts have to be considered.

Research in human geography may have similarities to research in other fields, but rarely will there be complete congruence, given the inherent logic of the disciplines. In other words, human geographers raise their own (new) questions, decide on specific field approaches, work on the basis of certain preferred methods and deal with very specific topics, possibly thereby repressing and/or neglecting some ethical issues too. A simple translation of findings from other disciplines into human geography is not possible without adjustments and, thus, translation will remain a permanent task. It is our hope that the present volume will help to master it.

References

Barker, J. and Smith, F. (2001): Power, positionality and practicality: Carrying out fieldwork with children. *Ethics, Place & Environment*, 4, 2, 142–147.

Barnett, C. (2011): Geography and ethics: Justice unbound. *Progress in Human Geography*, 35, 2, 246–255.

Barnett, C. (2012): Geography and ethics: Placing life in the space of reasons. *Progress in Human Geography*, 36, 3, 379–388.

Barnett, C. (2014): Geography and ethics III: From moral geographies to geographies of worth. *Progress in Human Geography*, 38, 1, 151–160.

Bell, K. (2014): Resisting commensurability. Against informed consent as an anthropological virtue. *American Anthropologist*, 116, 3, 511–522.

Boyd, W. B. E., Healey, R. L., Hardwick, S. W., Haigh, M., Klein, P., Doran, B., Trafford, J. and Bradbeer, J. (2008): 'None of us sets out to hurt people': The ethical geographer and geography curricula in higher education. *Journal of Geography in Higher Education*, 32, 1, 37–50.

Brant Castellano, M. (2014): Case 4: Research on Aboriginal People (pp. 273–288). In: Teays, W, Gordon, J.-S. and Renteln, A. D. (eds): *Global Bioethics and Human Rights. Contemporary Issues*. Lanham: Rowman & Littlefield Publishers.

Buller, H. (2016) Animal geographies III: Ethics. *Progress in Human Geography*, 40, 3, 422–430.

Burgess, M. M. (2007): Proposing modesty for informed consent. *Social Science & Medicine*, 65, 11, 2284–2295.

Cahill, C., Sultana, F. and Pain, R. (2007): Participatory ethics: Politics, practices, institutions. *ACME: An International Journal for Critical Geographies*, 6, 3, 304–318.

Coombes, B., Johnson, J. T. and Howitt, R. (2014): Indigenous geographies III: Methodological innovation and the unsettling of participatory research. *Progress in Human Geography*, 38, 6, 845–854.

Cutchin, M. P. (2002): Ethics and geography: Continuity and emerging syntheses. *Progress in Human Geography*, 26, 5, 656–664.

Dickson, A. and Holland, K. (2017): Hysterical inquiry and autoethnography: A Lacanian alternative to institutionalized ethical commandments. *Current Sociology*, 65, 1, 133–148.

Dyer, S. and Demeritt, D. (2009): Un-ethical review? Why it is wrong to apply the medical model of research governance to human geography. *Progress in Human Geography*, 33, 1, 46–64.

England, K. V. (1994): Getting personal: Reflexivity, positionality, and feminist research. *The Professional Geographer*, 46, 1, 80–89.

Eyles, J. and Smith, D. M. (eds). (1988): *Qualitative Methods in Human Geography*, Oxford: Blackwell.

Flicker, S. and Worthington, C. A. (2012): Public health research involving aboriginal peoples. Research ethics board stakeholders' reflections on ethics principles and research processes. *Canadian Journal of Public Health*, 103, 1, 19–22.

Fossheim, H. and Ingierd, H. (eds). (2015): *Internet Research Ethics*. Oslo: Cappelen Damm Akademisk.

Gannon, W. L. (2014) Integrating research ethics with graduate education in geography. *Journal of Geography in Higher Education*, 38, 4, 481–499.

Gaudet, S. and Robert, D. (2018): *A Journey Through Qualitative Research. From Design to Reporting*. Los Angeles, London, New Delhi: Sage.

Gill, R. (2017): Beyond Individualism. The Psychosocial Life of the Neoliberal University (pp. 193–216). In: M. Spooner (ed): *A Critical Guide to Higher Education & the Politics of Evidence. Resisting Colonialism, Neoliberalism, & Audit Culture*. Regina: University of Regina Press.

Gormley, N. and Bondi, L. (1999): Ethical Issues in Practical Contexts (pp. 251–262). In: Proctor, J. D. and Smith, D. M. (eds): *Geography and Ethics: Journeys in a Moral Terrain*. London: Routledge.

Guillemin, M. and Gillam, L. (2004): Ethics, reflexivity, and "ethically important moments" in research. *Qualitative Inquiry*, 10, 2, 261–280.

Happ, D., Meyer, F., Miggelbrink, J. and Beurskens, K. (2018): (Un-)informed Consent? Regulating and Managing Fieldwork Encounters in Practice (pp. 19–36). In: Wintzer, J. (ed): *Sozialraum erforschen: Qualitative Methoden in der Geographie*. Berlin, Heidelberg: Springer Spektrum.

Hay, I. (1998a): From code to conduct: Professional ethics in New Zealand geography. *New Zealand Geographer*, 54, 2, 21–27.

Hay, I. (1998b): Making moral imaginations: Research ethics, pedagogy, and professional human geography. *Ethics, Place & Environment*, 1, 1, 55–75.

Heimer, C. A. and Petty, J. (2010): Bureaucratic ethics: IRBs and the legal regulation of human subjects research. *Annual Review of Law and Social Science*, 6, 1, 601–626.

Herman, T. and Mattingly, D. J. (1999): Community, Justice and the Ethics of Research. Negotiating Reciprocal Research Relations (pp. 209–222). In: Proctor, J. D. and Smith, D. M. (eds): *Geography and Ethics: Journeys in a Moral Terrain*. London: Routledge.

Hopkins, P. E. (2007): Positionalities and knowledge: Negotiating ethics in practice. *ACME: An International Journal for Critical Geographies*, 6, 3, 386–394.

Horton, J. (2008): A 'sense of failure'? Everydayness and research ethics. *Children's Geographies*, 6, 4, 363–383.

Horton, J. (2020): Failure failure failure failure failure failure: Six types of failure within the neoliberal academy. *Emotion, Space and Society*, 35, 1–6.

Irwin, S. (2013): Qualitative secondary data analysis: Ethics, epistemology and context. *Progress in Development Studies*, 13, 4, 295–306.

Kearns, R. (1998): Interactive ethics: Developing understanding of the social relations of research. *Journal of Geography in Higher Education*, 22, 3, 297–310.

Kindon, S., Pain, R. and Kesby, M. (eds). (2007): *Participatory Action Research Approaches and Methods: Connecting People, Participation and Place*. London and New York: Routledge.

Kirman, J. M. (2003): Transformative geography: Ethics and action in elementary and secondary geography education. *Journal of Geography*, 102, 3, 93–98.

Lynn, W. S. (1998): Animals, Ethics and Geography (pp. 280–298). In: Wolch, J. and Emel, J. (eds): *Animal Geographies: Place, Politics and Identity in the Nature-Culture Borderlands*. London: Verso.

Maar, M. A., Sutherland, M. and McGregor, L. (2007): A regional model for ethical engagement. The first nations research ethics committee on Manitoulin Island. *Aboriginal Policy Research Consortium International (APRCi)*, 112, 55–68.

Madge, C. (2007): Developing a geographers' agenda for online research ethics. *Progress in Human Geography*, 31, 5, 654–674.

Manzi, M., Ojeda, D. and Hawkins, R. (2019): 'Enough wandering around!': Life trajectories, mobility, and place making in neoliberal academia. *The Professional Geographer*, 71, 2, 355–363.

McDowell, L. (1992): Doing gender: Feminism, feminists and research methods in human geography. *Transactions of the Institute of British Geographers*, 17, 4, 399–416.

McEwan, C. and Goodman, M. K. (2010): Place geography and the ethics of care: Introductory remarks on the geographies of ethics, responsibility and care. *Ethics, Place and Environment*, 13, 2, 103–112.

Moore, N. (2012): The politics and ethics of naming: Questioning anonymisation in (archival) research. *International Journal of Social Research Methodology*, 15, 4, 331–340.

Moreno, M. A., Goniu, N., Moreno, P. S. and Diekema, D. (2013): Ethics of social media research: Common concerns and practical considerations. *Cyberpsychology, Behavior and Social Networking*, 16, 9, 708–713.

Oberhauser, A. M. and Caretta, M. A. (2019): Mentoring early career women geographers in the neoliberal academy: Dialogue, reflexivity, and ethics of care. *Geografiska Annaler: Series B, Human Geography*, 101, 1, 56–67.

Olson, E. (2015): Geography and ethics I: Waiting and urgency. *Progress in Human Geography*, 39, 4, 517–526.

Olson, E. (2016): Geography and ethics II: Emotions and morality. *Progress in Human Geography*, 40, 6, 830–838.

Olson, E. (2018): Geography and ethics III: Whither the next moral turn? *Progress in Human Geography*, 42, 6, 937–948.

Olson, E. and Sayer, A. (2009): Radical geography and its critical standpoints: Embracing the normative. *Antipode*, 41, 1, 180–198.

Pain, R. (2004): Social geography: Participatory research. *Progress in Human Geography*, 28, 5, 652–663.

Popke, J. (2006): Geography and ethics: Everyday mediations through care and consumption. *Progress in Human Geography*, 30, 4, 504–512.

Popke, J. (2009): Geography and ethics: Non-representational encounters, collective responsibility and economic difference. *Progress in Human Geography*, 33, 1, 81–90.

Proctor, J. D. (1998): Ethics in geography: Giving moral form to the geographical imagination. *Area*, 30, 1, 8–18.

Proctor, J. (1999): Introduction (pp. 1–16). In: Proctor, J. D. and Smith, D. M. (eds): *Geography and Ethics: Journeys in a Moral Terrain*. London: Routledge.

Puāwai Collective. (2019): Assembling disruptive practice in the neoliberal university: An ethics of care. *Geografiska Annaler: Series B, Human Geography*, 101, 1, 33–43.

Rat für Sozial- und Wirtschaftsdaten. (2017): *Forschungsethische Grundsätze und Prüfverfahren in den Sozial- und Wirtschaftswissenschaften*. Berlin: Rat für Sozial- und Wirtschaftsdaten.

Richardson, J. C. and Godfrey, B. S. (2003): Towards ethical practice in the use of archived transcripted interviews. *International Journal of Social Research Methodology*, 6, 4, 347–355.

Rogers, R. (2019): *Doing Digital Methods*. London: Sage.

Rose, G. (1997): Situating knowledges: Positionality, reflexivities and other tactics. *Progress in Human Geography*, 21, 3, 305–320.

Rundstrom, R. and Deur, D. (1999): Reciprocal appropriation. Toward an ethics of cross-cultural research (pp. 237–250). In: Proctor, J. D. and Smith, D. M. (eds): *Geography and Ethics: Journeys in a Moral Terrain*. London: Routledge.

Ryen, A. (2010): Ethical Issues (pp. 218–235). In: Seale, C., Gobo, G. and Gubrium, J. F. (eds): *Qualitative Research Practice*. London: Sage.

Silverman, D. (2017): *Doing Qualitative Research*. 5th Edition. London: Sage.

Skloot, R. (2010): *The Immortal Life of Henrietta Lacks*. New York: Crown Publishers.

Smith, D. M. (1997): Geography and ethics: A moral turn? *Progress in Human Geography*, 21, 4, 583–590.

Smith, D. M. (1998): Geography and moral philosophy: Some common ground. *Ethics, Place and Environment*, 1, 1, 7–33.

Smith, D. M. (2001): Geography and ethics: Progress, or more of the same? *Progress in Human Geography*, 25, 2, 261–268.

Sultana, F. (2007): Reflexivity, positionality and participatory ethics: Negotiating fieldwork dilemmas in international research. *ACME: An International Journal for Critical Geographies*, 6, 3, 374–385.

Valentine, G. (2003): Geography and ethics: In pursuit of social justice, ethics and emotions in geographies of health and disability research. *Progress in Human Geography*, 27, 3, 375–380.

Valentine, G. (2004): Geography and ethics: Questions of considerability and activism in environmental ethics. *Progress in Human Geography*, 28, 2, 258–263.

Valentine, G. (2005): Geography and ethics: Moral geographies? Ethical commitment in research and teaching. *Progress in Human Geography*, 29, 4, 483–487.

Whatmore, S. (1997): Dissecting the autonomous self: Hybrid cartographies for a relational ethics. *Environment and Planning D: Society and Space*, 15, 1, 37–53.

White, C. and Bailey, C. (2004): Feminist knowledge and ethical concerns: Towards a geography of situated ethics. *Espace Populations Sociétés*, 1, 131–141.

Winchester, H. P. (1996): Ethical issues in interviewing as a research method in human geography. *The Australian Geographer*, 27, 1, 117–131.

Yin, R. K. (2018): *Case Study Research and Applications: Design and Methods.* 6th Edition. Los Angeles: Sage.

Zhang, J. J. (2017): Research ethics and ethical research: Some observations from the Global South. *Journal of Geography in Higher Education*, 41, 1, 147–154.

Part I

Ethics in human geographical research

Part 1

Ethics in human geographical research

2 Caring about research ethics and integrity in human geography

Iain Hay and Mark Israel

Introduction

Take a moment to imagine a world of human geographic inquiry in which researchers care little for ethics and integrity in their work. Just what would that be like? Would members of the public give us access to details of their lives and beliefs? How much harm might we do to research participants through a disregard for their well-being? Could we trust one another's work? Would we have faith in the veracity and accuracy of previous studies? Could we work confidently with colleagues? Would we even want to work with them? Could the public believe anything we reported to them? Would the government or other agencies see any value in endorsing and supporting our work? For many of us, the answer to each of these questions would be 'no'.

Over the past 20 years, there has been a rise in the volume of work on research ethics in the social sciences generally (Catungal and Dowloing 2021; Dingwall et al. 2017; Eriksen 2017; Hay 2013, 2016; Israel 2015; Proctor 1998; Smith 1997; Valentine 2005). However, very little of that published material deals in any detail with the question 'why care about research ethics in our particular discipline?' The answer appears to be taken as a given: it is a matter so obvious that no explanation is called for. Problematically for geographers, egregious, and high-profile examples of bad practice in other fields, such as biomedical research or psychology, for instance, have motivated widespread concern, and the development of codes of conduct that are now applied to human geography and other social sciences (for a discussion see Israel and Hay 2006 or Israel 2015).

In general, human geography has taken the rationale for its concern about research ethics from these other fields and paid less attention to the discipline's own foundations of interest. Indeed, at a disciplinary level, because there may be a view that major ethical calamities can't happen to us, we find ourselves swept along quietly on a rising tide of moral regulation, generated by challenges from other fields. While human geography has for the most part been spared high profile research horrors and humiliations, that relatively unruffled history may also have contributed to an arguable lack of conscious, authentic engagement with research ethics, taking them for granted, and aiming only to satisfy the procedural expectations of ethics

DOI: 10.4324/9780429507366-2

review committees, rather than embedding careful and thoughtful ethical conduct throughout everyday practice (Hay 1998). If we are to avoid the prospect of ungrounded and hollow ethical conduct and evade contributing to the dystopian research landscape signalled above, we need to care about research ethics, have a sound appreciation of why we care, make decisions based on ethical analysis, and then share our reasoning with our colleagues inside and outside our discipline.

In this chapter, we want to look at just one part of this agenda and explore some of the motivations for human geographers (like many other social scientists) to engage with ethics. Why should we care about ethics? We argue that, among other reasons, human geographers need to engage with ethics in order to minimise harm, increase the sum of good and pursue justice, cope with new and challenging methodological and social problems, and to assure public trust and promote broader moral and social values.

To protect others (and ourselves), minimise harm, and increase the sum of good

One of the most important and self-evident of the reasons human geographers need to care about research ethics is to ensure that through our work we protect others, minimise harm, and increase the overall sum of good. As social scientists perhaps trying to advance social justice by, for example, contributing to health and well-being, making neighbourhoods safer, improving children's lives and education, and guaranteeing democracy (Olser 2011), our work should protect individuals, communities, and environments from both short-term and enduring harm.

It would be comforting to think there is something intrinsic to the human condition that encourages such benevolent behaviour but sadly, the history of research challenges that view. Activities involving scandalously unethical conduct have been uncovered in the years since the end of the Second World War (Resnik 2015). High profile examples of these in the United States include abhorrent research conducted on prisoners before, during and after the Second World War (Israel 2016); the Tuskegee syphilis study (Reverby 2009); US government studies in Guatemala that intentionally infected over 1300 subjects with venereal diseases to test the effectiveness of penicillin (Presidential Commission for the Study of Bioethical Issues 2011); the Willowbrook State School hepatitis experiments (Robinson and Unruh 2008); CIA mind control research (McCoy 2006), which included administering LSD to unwitting subjects; the complicity of the American Psychological Association in a CIA torture scandal (Ackerman 2015; Risen 2015); and embedding social scientists within the US Army's Human Terrain System in Iraq and Afghanistan (American Anthropological Association 2009; Gregory 2015). Though drawn from fields other than human geography, these examples do point to social and personal matters that encourage, allow or condone ethically questionable work.

Of course, as human geographers we have faced our own challenges associated with protecting others, minimising harm and increasing the sum of good. One example is the American Geographical Society's 2005–2008 Bowman Expedition, 'México Indígena' in Oaxaca (Voosen 2016). This case involved US geographers, led by Peter H. Herlihy from the University of Kansas, mapping Mexico's vast neoliberal land certification and privatisation program (Herlihy 2010). The project worked towards a nationwide database of property rights and received significant US defence funding through the Foreign Military Studies Office. A key point of dispute was whether the researchers informed communities of the US military's role in the project (Bryan 2010). Scrutiny of the Bowman project sparked intense debate about the implications of military sponsoring and use of geographic research (see Agnew 2010; Bryan 2010; Cruz 2010; Herlihy 2010; Steinberg 2010; Steinberg et al. 2011) in threatening the autonomy of Indigenous peoples. Further examples of the ethics of 'everyday' human geographic work are described in Hay (2016) and Hay and Israel (2008). We have identified some others in Box 2.1 which can be used to explore situations where human geographers have to consider: how to respect or enhance the autonomy of people and communities; the protection of others; minimisation of harm, increasing the sum of good, and ensuring that any harm is justified by the benefit; and conflicts of interest and roles.

As some of the examples set out in Box 2.1 suggest researchers may also, and inadvertently, harm themselves. In insightful recent work, Australian geographer Christine Eriksen (2017) points to the vicarious and debilitating trauma she experienced over six years of ethnographic work with bushfire survivors, firefighters and residents fearful of bushfire threat. Despite the broad attention we give to the harmful effects of research, she concluded that academics typically fail to consider the mental (or other) distress that

---

Box 2.1

Examples of difficult ethical situations encountered by human geographers[1]

Case Study	Issue Covered
Tomotsugu is a visiting professor. For a book he is writing he wishes to map locations of historical indigenous sites. The mapping exercise could help government and other agencies protect sites from building and other harmful activities. However, Tomotsugu's plans go against the wishes of local indigenous peoples and could open up some sites to physical damage.	1 Mapping indigenous sites 2 Physical harm to cultural sites 3 Harm to indigenous community

[1]Except where cited, these examples have been redacted and are based on real cases provided to the first author on the assurance of anonymity.

A government space-saving and paper recycling scheme requires many government documents be routinely destroyed. One public servant, Ellis, who is also a PhD student has the opportunity to secretly hold back from this process a line of papers that may be relevant to her academic research. Singly, the documents have little value but when aggregated they provide a large body of evidence to support the argument in Ellis' PhD that public land-use planning and management is often driven by political expediency and commercial imperatives.

1 Dual roles/conflict of roles
2 Breaching confidentiality
3 Covert research
4 Public benefit

Nelson is in a foreign country on a research visit sponsored by the host nation and the US government. At an appointment with US Embassy officials. Nelson is asked to look for special information on local activities the US wishes to stop. Embassy officials point out to Nelson that her position as an academic researcher will allow her to move freely whereas embassy personnel are under constant surveillance. Although no one mentions it, Nelson is aware that this intelligence gathering exercise could threaten the lives of others. To refuse will mark her as uncooperative in the eyes of the US officials. Nelson also relies on the Embassy for information and assistance for her work.

1 Physical harm
2 Conflict of roles
3 Deception

Yee is conducting research for her thesis. In consultation with her supervisor, she abides by ethical guidelines laid down by her university. One of the standard requirements of the Research Ethics Committee (Institutional Review Board) is that researchers provide prospective research 'subjects' with a letter of introduction promising that 'any information provided will be treated in the strictest confidence and none of the participants will be individually identifiable in the resulting thesis'. Yee's work involves three of the four major cinema complexes in the city where her university is based. Her research involves semi-structured interviews with the three cinema managers. The material to be covered relates to the managers' professional roles and not their private lives. All three have agreed to participate in the study under the terms outlined. It is only now that Yee realises she will not be able to write her thesis in such a way that the anonymity of the cinema managers can be assured for even if she uses pseudonyms for managers and cinemas it will be easy for readers of the thesis to identify the individual respondents. Time is pressing. Margaret only has limited time left to complete the research that she has already been working on for six months and to write her thesis.

1 Anonymity
2 Informed consent

Song is a feminist geographer conducting research on the division of household labour in the suburbs of a large city. As part of her work, she is interviewing people involved in heterosexual relationships to explore the ways they regard and act on the division of household labour. Song endeavours to interview each member of each couple apart from her/his partner but, in some instances, she finds this is not possible. Some interviews with couples become strained as partners hear and react to one another's thoughts on the ways their household work is divided.

1 Emotional harm
2 Confidentiality

Urban geographer Bradley Garrett was arrested in London in 2012. Four years earlier he had started doctoral research on 'urban explorers' or 'place hackers'. Garrett's exploration of British urban space involved trespass onto land owned by the public transport authority. Garrett and eight of his research participants were charged with conspiracy to commit criminal damage. The prosecution argued Garrett's law-breaking was both unethical and unnecessary since he could have completed the work legally. They might also have pointed to the possibility that repeated trespass would have required the authority to spend more on security. In turn, the defence drew on experts who indicated that if ethnography were to be 'deep and full' it might well require engagement in interactions and situations that are illegal (Ferrell et al. 2015). Garrett himself argued that it was 'deeply problematic' to block research by people simply because they lived close to 'legal boundaries'. He also argued that participant observation with such groups might entail breaking the law. The case ended with Garrett (2014) receiving a conditional discharge.	1 Physical harm 2 Legal harm 3 Illegal activity

may befall researchers. And so, she sagely advocates 'an ethical framework attentive to researcher mental health' (Eriksen 2017, p. 275):

> Participant well-being, including mental health, is already high on the list of concerns of the HREC [Human Research Ethics Committee]. HRECs could take the cognitive and progressive leap towards a similar concern for the mental health of researchers ....
>
> (p. 276)

Eriksen's experience, the Oaxaca incident, and the examples provided in Box 2.1 point to ways human geographic work has the potential to be damaging and problematic – both to participants and researchers. Given the juxtaposition of challenging situations against conditions that can encourage, allow or condone questionable work, the need seems clear for continuing individual professional reflection on our own decisions and conduct, critical – yet collegial – consideration of the work of our colleagues, and collective efforts to build the capacity of our discipline to engage in ethical analysis and act on the consequences of this analysis.

To cope with new and more challenging problems and methods

It is not uncommon for human geographers to face ethical challenges and dilemmas in their research. Over the past decade, rapid growth in the interest of human geographers in socially, politically and economically marginalised groups of people (von Benzon and van Blerk 2017) together with the growing methodological and technological complexity of human geographical research has begun to raise more and more demanding ethical issues. Emphases on cross-cultural, Indigenous, and participant action research in geography (e.g. Gergan and Smith 2021; Howitt and Stevens 2016; Johnson and Madge 2021; Kindon 2021; Tobias et al. 2013; Yanar et al. 2016) as well as geographers' activity in interdisciplinary projects present significant questions about researcher roles and positions. Researching the experiences of socially, politically and economically marginalised groups of people, is not a new endeavour. What is new, however, is the rapidity in the growth of interest of researchers seeking to engage with these populations, and the variety of the toolkit of methods available to support this. For instance, researchers may track the movement of marginalised individuals through urban space in low-income countries, use the geotags on Twitter feeds to assess the evacuation responses of residents facing catastrophic events (Martín et al. 2017) or create heat maps of energy use to plan and construct technologically smart energy-efficient houses. Scholars have had to deal with ethical issues such as gender-related levels of personal disclosure; relationships between 'readability' and fidelity; benefit sharing; anonymity, confidentiality and big data; and, questions of authenticity and credibility (Harrison and Lyon 1993; Israel 2015). Likewise, internet-based research continues to raise extensive issues surrounding personal privacy and public observation in cyberspace (Markham and Buchanan 2012).

New geospatial technologies also present challenges. These can compromise an individual's locational privacy (mobile phones are an everyday example, discussed recently by Taylor 2015). Even when masked deliberately in research findings, an individual's location can be uncovered with the aid of geospatial technologies – by people with little knowledge of those technologies (Kounadi and Leitner 2014). Indeed, in one case researchers were able to identify up to 79 per cent of patient addresses in hypothetical health geography research employing maps with patient locations masked (Brownstein et al. 2006). They concluded that the publication of low-resolution disease maps has the potential to jeopardise patient privacy, accidentally disclosing patient information, and therefore we need guidelines for the publication of health data in order to help guarantee patient privacy. Indeed, human geographers need to take the lead on advising how to handle geographically identifiable personal data and when it might be ethically acceptable to link such data to that typically used by disciplines without a long history of social research such as computer science, engineering and agricultural science. Some members of these disciplines are well aware that they also need the support of social scientists to understand how technological innovation might have a social impact (Gibney 2018) and several guidelines on the ethics of using spatial data have been developed in the last few years (see Berman et al.; UK Statistics Authority 2021; W3C Interest Group 2021).

Regrettably, institutional codes of ethics, regulations and associated educational materials may offer very little support to researchers grappling with such emerging research challenges (Allen and Israel 2018). So, as responsible practitioners concerned about research ethics, we need to give careful thought to the complex and unique dilemmas we encounter and make informed decisions about our work that we can defend to ourselves and critical others.

Because we have not really contributed to research ethics debates and education

The more challenging problems and methods confronting (human) geographers introduced above highlight the need for us to develop and contribute to a better understanding of ethics – from principles to practices – and to ensure too that our students are skilled and knowledgeable actors in these domains. Moreover, and importantly, because of the scope of our work, geographers can be especially well-placed to make effective contributions to ethical debate. For instance, a number of key ethical principles are held to underpin approaches to ethics developed at institutional, national and supranational levels associated with research involving human participants. This principlism, as it is known, has its formal origins in the Belmont Report (NCPHSBBR 1979) in the United States, now an internationally influential statement of ethical principles and guidelines intended to assist in resolving ethical problems. Principlism draws on both consequentialist and deontological traditions, the key elements of European and Western post-Enlightenment ethics. The principles require we demonstrate respect for all persons involved in our work; exhibit beneficence or at least non-maleficence; and act in ways that can be regarded as just. Principlism has its critics who argue that, among other matters, it lacks adequate foundation in theory, promotes values and approaches to knowledge and method that privilege the Global North (Morris 2015), favours individual autonomy over communal relationships, ignores a wide range of other ethical approaches such as Indigenous, Feminist and critical approaches (Denzin et al. 2008) and frustrates attempts to link ethics to broader questions of social justice (Hammersley 2013; Israel 2015). Clearly, these are all debates to which human geography can and should contribute. For example, geographers can make a unique contribution to tracing international policy transfers and the spatial configurations of distributive justice.

There are dangers for us if we do not contribute. Social scientists have complained that particular methodologies such as covert research (Calvey 2017) and autoethnography (Tolich 2010) are being blocked by research ethics review; others have noted that the review process is insufficiently swift to allow students engaged in short research projects to pass review and undertake fieldwork (Bledsoe et al. 2007). Students have been steered away from emerging and in-depth qualitative research before they start research higher degrees (van den Hoonaard 2011). The universalising tendency of principlism and those policy communities that support its extension across disciplines and countries (Israel 2018) ought to be countered by disciplines like human geography that are sensitive to the spatial specificity of cultural and bureaucratic practices.

And yet, despite abundant institutional rhetoric surrounding the value with which ethical conduct is held, with few exceptions (e.g. Gannon 2014; Vujakovic and Bullard 2001) ethics receives little comprehensive attention in many geography higher education courses (Healey and Ribchester 2016, pp. 302–303; van den Bemt et al. 2018; Zhang 2017;). In short, many of us have graduated from educational systems that have failed to support our understanding of ethics and our capacity to manage ethical dilemmas and challenges. Indeed, for many students and early career researchers their main understanding of ethical research draws from mechanistic and limited-term encounters with institutional ethics clearance procedures (Boyd et al. 2008; Kearns 1998). So, for example, despite their obvious value helping geographers negotiate local institutional review processes, discipline-specific research ethics resources such as at Durham University (2017a, b) and the University of Exeter (2018) in the United Kingdom appear to do little to support more universal sensitivity to, and analysis of, ethical challenges. We argue it is incumbent on us as researchers independently and collectively to explore the underpinnings, practice, and significance of ethical research and to use this to nurture the moral imaginations of a new generation of human geographers (Hay 1998).

To assure public trust and support for our professional work

We also need to care about research ethics as part of the professionalisation of human geography and its engagement with the broader community. As Marcuse (1985, p. 20) has noted, professionalisation has played a historical role in sealing 'a bargain between members of the profession and the society in which its members work'. And so, as part of claims to professional status or recognition for members – as well as to ensure virtuous research – bodies such as the American Association of Geographers, the United States' National Geographic Society, the New Zealand Geographical Society, and the Royal Geographical Society (with the Institute of British Geographers) have adopted processes and procedures for self-regulation of members' ethical conduct. In return, members claim and accept professional status and associated privileges. These benefits include the public credibility and trust that can facilitate the conduct of research with particular individuals and communities. Sadly, however:

> The rise in litigation between human subjects and the research establishment suggests that there is a growing loss of trust between investigators and human subjects. This must be reversed, and to do so it is essential that investigators establish relationships based on trust and respect with their research subjects and maintain the highest ethical standards.
>
> (Yale University 2018)

Ethical research reflects the respect we have for the people who agree to participate in our work. It also helps build and sustain support for our scholarship. Because we have no inalienable right to conduct research involving other people, species, or places our continuing freedom and scope to work is the product of individual and social goodwill and depends on behaviours that are respectful, beneficent and just. Where prior trust has been violated – with, for example, Pacific Island communities (Walsh 1992), Indigenous peoples (Howitt and Stevens 2016; Lavallée and Leslie 2016) or 'vulnerable' groups (von Benzon and van Blerk 2017) – we may have to work hard and long to regain public support.

Greater care with respect to research ethics may help assure a climate of trust in which we can undertake socially useful labours. If geographers act in ways that demonstrate respect for others, beneficence, trust, justice and fidelity, people may rely on us to recognise their needs and sensitivities and consequently be more willing to contribute openly and fully to the work we undertake.

The merit of our scholarly work depends on it being undertaken ethically and disseminated honestly. Yet, in much of our work, there is no one present on an everyday basis to oversee whether we are doing the 'right' thing or not. A consequence of geographers' highly valued research autonomy is that for the most part we are left to – and must – control our own conduct. This takes heightened significance as pressures on academic integrity increase on the back of growing dependence of universities and researchers on sponsorship and linking government grants, institutional pressures to climb international rankings, as well as to promotions and salary increments linked to research 'performance' (Fong and Wilhite 2017; Necker 2014). The need for self-regulation is not to deny the legitimate need for parallel forms of ethics governance (see Sheehan et al. 2018) but, in order to support and promote the integrity of our research, we must care about ethics genuinely and act in ways that accord with that care – even when no-one is watching.

Ensuring the ethical conduct of collaborators is important to us both as individual scholars and collectively. Because so much of our research depends on and builds on other's findings and advances, if those contributions are fabricated or dishonest or obtained through questionable means, it is we who bear many of the costs (e.g. through research built on foundations that are incorrect or untrue). So, as individuals drawing from colleagues' work, it is in our self-interest – as well as the interests of the discipline and community as a whole – to care deeply about research ethics in our own practice as well as in the practices of our colleagues. It bears remembering that 'While often incredible in hindsight, the feats of all of the perpetrators of [known] scientific misconduct have happened under the (watchful) eyes of their subordinates, peers and superiors' (Malički et al. 2017, p. 1).

The number of papers that have been retracted by journals on grounds of misconduct – often on the basis of plagiarism, fabrication and falsification – has risen considerably over the last two decades (Fanelli 2014).

There are few examples involving geographers, but there are some. In 2010, *Global Ecology and Biogeography* retracted a paper on mid-Holocene forest decline in China, according to *Retraction Watch* as a result of questions about authorship (Oransky 2010). A 2015 article on 'Early hominin biogeography in Island Southeast Asia', was retracted by *Evolutionary Anthropology*, following complaints that the authors had referred to personal correspondence without the permission of the corresponding party (Marcus 2017). Most seriously, in 2014, two Springer journals found that the peer review process relating to two articles on environmental change and environmental management in China by authors from the School of Geography at Beijing Normal University had been compromised by the first author acting as a favourable peer reviewer of the articles (Oransky 2014). In 2013, *Antipode* retracted an article on sexuality and geography where there had been 'unattributed prior publication of a substantially similar version' in a magazine in Dutch. In fairness to the author, there might be reasons for publishing similar material in different contexts and languages and this is something that we ourselves have contemplated; however, the fault appears to be that in this case this practice was not acknowledged. This final case is not as clear cut as the others and provides a good example of why research ethics and integrity require the development of an ethical or 'moral imagination' (Hay 1998) rather than the creation and application of rules.

To satisfy organisational and professional demands

High profile and flagrant examples of scientific misconduct and impropriety together with a rapidly growing international emphasis on thorough corporate governance (Solomon 2007), have driven overlapping public and institutional demands that collectively underpin further self-interested reasons for human geographers to care about research ethics. Many commentators have written about how regulatory requirements have intensified as well as being extended to additional jurisdictions, disciplines and methodologies (Israel 2018; Schrag 2010).

Not surprisingly, in many parts of the world universities, funding agencies, employers, and professional and scholarly societies seek to protect themselves from damaging reputational and legal consequences of unethical actions of employees, members, or representatives. And to ensure they are protected – as well as to consolidate their own public profiles as caring, community-minded organisations – these agencies mandate that researchers take ethics seriously, often with grave consequences for failure to comply. Ramnath et al. (2016) describe the various corrective actions that US institutional review boards have taken when researchers have failed to comply with research ethics approval conditions. There are also a few older examples where the Office for Human Research Protections (2001) has suspended research in United States institutions, though none of the failures have been connected to geographers. The spectres of funding withdrawal, legal action and poor publicity drive careful monitoring and review of (un) ethical practices as part of broader institutional risk management strategies. While such oversight may yield better, more humane research, its emphasis has been more focussed on regulatory compliance than on ethical conduct (Bledsoe et al. 2007; Israel 2015). Increased institutional oversight demands we 'care' about ethics, or risk facing greater and more intrusive external surveillance of our conduct.

Human geographers also need to care about research ethics to minimise risks of personal liability. Institutions commonly distribute their corporate insurance responsibilities for ensuring ethical research conduct by demanding compliance from employees, researchers and students. As Western and other societies become more litigious, indemnity and insurance are becoming increasingly important for research organisations, with some being bound to ensure researchers sanctioned by the institution have satisfied all relevant ethics review requirements, as well as other less relevant ones such as a misplaced page in a consent form (see Bledsoe et al. 2007, p. 608). In effect, research institutions are brought into line by insurers and seek to manage their own risks by dissociating themselves from non-compliant employees, contractors and students.

Since a broader society provides resources that support many of our investigations, we need to care about research ethics to ensure funds are spent in ways that satisfy public demands and expectations. This powerful rhetoric of public accountability is used by institutions and organisations to justify all manner of governance and oversight arrangements (Suspitsyna 2010). However, as Sheehan et al. (2018, p. 3) note, 'society does not always provide such resources. These resources might come from private donations or from charitable trusts. [And], when society does provide the resources for research, it may be paying for the system that manages the research rather than for particular research projects. In addition, in the face of diminishing public funding for the humanities, arts and social sciences, in particular, funding from the corporate sector has become increasingly important. Regardless, the rhetoric and practices of public accountability are pervasive. What basis is there to suggest for example that the strictures and structures that surround publicly supported work – whether it be the actual conduct or management of that work – should not also be offered to private contributors or charities? And over and above any university or institutional regulation of privately derived funds, the agencies providing those funds have their own imperatives to monitor the conduct of grantees and provide advice to do so.

Increasingly, we need to care about research ethics if we wish to be published. Organisations like the Committee on Publication Ethics (COPE), established in 1997, now guide publishers and editors on systematic and comprehensive ways of minimising and managing misconduct (see COPE 2018). Their core practices on ethical oversight, intellectual property, and misconduct provide editors with well-defined and consistent guidance on research ethics matters and are helping raise the bar for authors wishing to publish in reputable journals. For example, growing numbers of journals will not accept for publication papers reporting research that has not been approved by an appropriate ethics committee. In many cases, the requirement started among biomedical journals; adopted by large publishers, the demand was extended across all their journals. This is unsurprising given revelations of misconduct in research – mostly outside social sciences. However, some geographers will struggle to meet the requirement that they obtain formal ethics clearance as, outside North America, Australasia, and the United Kingdom, the vast majority of countries do not mandate review of human geography research by universities. So, in many institutions governance structures are not in place that could offer the sign-off sought by some journals; outside universities, geographers working freelance, or in government departments and non-government organisations may have even less chance of finding a review committee. Editors of geography journals need to be aware of the consequences of adopting such a contextually insensitive policy, and clearly, some are, even at the risk of contravening their publisher's requirements (Robson 2018).

To respond to drifting moral anchors

At the same time as geographers are confronted by new methodological and technological challenges, traditional religious and other sources of moral justification and guidance are, for many people, decreasing in authority (Žižek 2000). The 'blame' for drifting moral anchors can be associated with all manner of social change and events, from liberalism, to perceived inconsistencies and contradictions in climate science, food sciences and medical sciences, and the demise of nuclear families, extended families, religious authority and local communities with corresponding declines in civic life and engagement. Arguably, widespread decentring of moral and professional authority has been joined with the mainstreaming of postmodernist critique which, at its core, casts doubt on authoritative definitions and uniform narratives of events. Some influential scholars such as Zygmunt Bauman (1993) have gone so far as to suggest that, through its 'incredulity towards metanarratives', postmodernity has dashed notions of universal, solidly grounded ethics regulation. Emerging from this confluence of social events and philosophical transformation are heightening pressures on the legitimacy of any individual or institution's claims to moral authority. Where once social scientists may have had the inherent trust and respect of the communities within which they work, that authority and legitimacy is made increasingly vulnerable by the heightened public profile of biomedical, social, political and environmental controversies and scientific debacles as well as by the conjunction of increasing education levels and heightening awareness of ethical matters among potential research participants. Authority and respect must now be earned, upheld and defended. Indeed, Donald A. Brown (2002, p. vii) observes:

> Moral authority cannot ... emanate from the top downward in any social structure. Instead, it must of necessity emerge from the bottom up. This is so because moral authority rests ultimately on a widely shared and mutually accepted sense of legitimacy.

So, efforts to maintain our credibility and social licence to conduct and continue our work have become another of the central reasons human geographers must carry on caring about research ethics.

And finally, on this matter of relativism, none of us is immune to broader ethical dislocations and uncertainties. Just as other groups question the foundations and meaning of virtue, so do we as human geographers. And so, how are we to know what is right behaviour? On what bases do we make such determinations? In an era of drifting moral anchors, each of us needs to care about research ethics, becoming skilled and independent ethical thinkers and actors, engaging actively with the meaning and practice of virtuous research, explaining our decisions and defending them openly.

Conclusion

The conduct of our research is involved in a mutually constitutive relationship with broader social processes, with each being reflective and supportive of the other's fundamental qualities. Thus, 'many of the norms of research promote a variety of other important moral and social values, such as social responsibility, human rights, animal welfare, compliance with the law, and public health and safety' (Resnik 2015, para. 11). How we conduct our research activities contributes to the moral shape of the societies of which we are a part. A microcosm of this demonstration effect can be seen in the way many research students learn about ethical conduct through the example of their mentors and supervisors. Moreover, and importantly, because social scientists continue to be held in high regard socially (Funk and Kennedy 2017; Lamberts et al. 2010), so our conduct may be held as a benchmark for the behaviour of others.

In the introduction to this chapter, we foreshadowed the dystopian landscape of human geographic inquiry that might emerge if we fail to care about ethics. It would be a world in which we could not trust one another's work. We could have no faith in the accuracy of other studies. We could not work confidently with colleagues. Governments and other agencies would cease to support our work. Harm to participants would drive them away from us.

We need to care deeply about research ethics. We need to care because although we may know how to negotiate institutional ethics review, we have not really learned about ethics. We need to care if we are to conceive what it would mean to make the world a better place and then act on the basis of that analysis. We need to care to ensure continuing public trust and support and to warrant that our own work, and that of our colleagues, is of the highest standard. We need to care if we are to satisfy professional and organisational demands that protect those institutions. We need to care because we are spending other people's money and because we want to be published. We need to care because every day new technologies and dynamic socio-spatial relationships throw up novel and more troubling problems to challenge us. And we need to care because many taken-for-granted moral anchors that may once have offered guidance and structure have been brought into question, demanding that we develop new capabilities to identify, support and engage in ethically defensible behaviour.

References

Ackerman, S. (2015) 'Three senior officials lose their jobs at APA after US torture scandal', https://www.theguardian.com/us-news/2015/jul/14/apa-senior-officials-torture-report-cia (accessed 28 January 2019).

Agnew, J. (2010) 'Ethics or militarism? The role of the AAG in what was originally a dispute over informed consent', *Political Geography*, 29 (8): 422–423.

Allen, G. and Israel, M. (2018) 'Moving beyond regulatory compliance: Building institutional support for ethical reflection in research', in R. Iphofen and M. Tolich, (eds) *The SAGE Handbook of Qualitative Research Ethics*, London: Sage, pp. 276–288.

American Anthropological Association (2009) Commission on the Engagement of Anthropology with the US Security and Intelligence Communities (CEAUSSIC) 'Final Report on the Army's Human Terrain System Proof of Concept Program', https://s3.amazonaws.com/rdcms-aaa/files/production/public/FileDownloads/pdfs/cmtes/commissions/CEAUSSIC/upload/CEAUSSIC_HTS_Final_Report.pdf (accessed 02 February 2019).

Bauman, Z. (1993) *Postmodern Ethics*. Oxford: Blackwell.

Berman, G., de la Rosa, S., and Accone, T. (2018) 'Ethical Considerations When Using Geospatial Technologies for Evidence Generation', Innocenti Discussion Papers no. 2018-02, UNICEF Office of Research - Innocenti, Florence. https://www.unicef-irc.org/publications/971-ethical-considerations-when-using-geospatial-technologies-for-evidence-generation.html (accessed 19 July 2021).

Bledsoe, C.H., Sherin, B.L., Galinsky, A.G., Headley, N.M., Heimer, C.A., Kjeldgaard, E., Lindgren, J.T., Miller, J.D., Roloff, M.E., and Uttal, D.H. (2007) 'Regulating creativity: Research and survival in the IRB iron cage', *Northwestern University Law Review*, 101 (2): 593–641.

Boyd, W., Healey, R.L., Hardwick, S.W., Haigh, M., Klein, P., Doran, B., Trafford, J., and Bradbeer, J. (2008) "None of us sets out to hurt people': The ethical geographer and geography curricula in higher education', *Journal of Geography in Higher Education*, 32 (1): 37–50.

Brown, D.A. (2002) *American Heat: Ethical Problems with the United States', Response to Global Warming.* Maryland: Rowman & Littlefield Publishers.

Brownstein, J.S., Cassa, C.A., Kohane, I.S., and Mandl, K.D. (2006) 'An unsupervised classification method for inferring original case locations from low-resolution disease maps', *International Journal of Health Geographics*, 5: 56.

Bryan, J. (2010) 'Force multipliers: Geography, militarism, and the Bowman expeditions', *Political Geography*, 29 (8): 414–416.

Calvey, D. (2017) *Covert Research: The Art, Politics and Ethics of Undercover Fieldwork*, London: Sage.

Catungal, J.P. and Dowling, R. (2021) 'Power, subjectivity, and ethics in qualitative research', in I. Hay and M. Cope, (eds) *Qualitative Research Methods in Human Geography*, 5th edition, Don Mills, Canada: Oxford University Press, pp. 18–39.

Committee on Publication Ethics (COPE) (2018) 'Committee on Publication Ethics (COPE)', https://publicationethics.org/ (accessed 02 February 2019).

Cruz, M. (2010) 'A living space: The relationship between land and property in the community', *Political Geography*, 29 (8): 420–421.

Denzin, N.K., Lincoln, Y.S., and Smith, L.T. (eds) (2008) *Handbook of Critical and Indigenous Methodologies*, Los Angeles: Sage, pp. 1–20.

Dingwall, R., Iphofen, R., Lewis, J., Oates, J., and Emmerich, N. (2017) 'Towards common principles for social science research ethics: A discussion document for the academy of social sciences', in R. Iphofen (ed) *Finding Common Ground: Consensus in Research Ethics Across the Social Sciences*, Bingley: Emerald Publishing Limited, pp. 111–123.

Durham University (2017a) 'Geography research ethics', https://www.dur.ac.uk/geography/research/ethics/ (accessed 02 February 2019).

Durham University (2017b) 'Department of Geography', https://www.dur.ac.uk/geography/research/ (accessed 28 January 2019).

Eriksen, C. (2017) 'Research ethics, trauma and self-care: Reflections on disaster geographies', *Australian Geographer*, 48 (2): 273–278.

Fanelli, D. (2014) 'Rise in retractions is a signal of integrity', *Nature*, 509: 33. doi.org/10.1038/509033a.

Ferrell, J., Hayward, K., and Young, J. (2015) *Cultural Criminology*, 2nd edition, London: Sage.

Fong, E.A. and Wilhite, A.W. (2017) 'Authorship and citation manipulation in academic research', *PLOS ONE*, 12 (12). doi.org/10.1371/journal.pone.0187394.

Funk, C. and Kennedy, B. (2017) 'Public confidence in scientists has remained stable for decades', http://www.pewresearch.org/fact-tank/2017/04/06/public-confidence-in-scientists-has-remained-stable-for-decades/ (accessed 02 February 2019).

Gannon, W.L. (2014) 'Integrating research ethics with graduate education in geography', *Journal of Geography in Higher Education*, 38 (4): 481–499.

Garrett, B. (2014) 'Place-hacker Bradley Garrett: Research at the edge of the law', https://www.timeshighereducation.com/features/place-hacker-bradley-garrett-research-at-the-edge-of-the-law/2013717.article (accessed 02 February 2019).

Gergan, M. and Smith, S. (2021) 'Reaching out: Cross-cultural research', in I. Hay and M. Cope, (eds) *Qualitative Research Methods in Human Geography*, 5th edition, Don Mills, Canada: Oxford University Press, pp. 40–59.

Gibney, E. (2018) 'The ethics of computer science: this researcher has a controversial proposal', *Nature*. doi.org/10.1038/d41586-018-05791-w.

Gregory, D. (2015) '(In)human terrain geographical imaginations', https://geographicalimaginations.com/2015/08/04/inhuman-terrain/ (accessed 28 January 2019).

Hammersley, M. (2013) 'A response to 'Generic ethics principles in social science research' by David Carpenter', http://acss.org.uk/wp-content/uploads/2014/01/Hammersley-AcSS-Response-to-Carpenter-5-March-2013-Principles-for-Generic-Ethics-Principles-in-Social-Science-Research.pdf (accessed 02 February 2019).

Harrison, B. and Lyon, E.S. (1993) 'A note on ethical issues in the use of autobiography in sociological research', *Sociology*, 27 (1): 101–109.

Hay, I. (1998) 'Making moral imaginations. Research ethics, pedagogy, and professional human geography', *Philosophy & Geography*, 1 (1): 55–76.

Hay, I. (2013) 'Geography and ethics', in B. Warf (ed) *Oxford Bibliographies in Geography*, New York: Oxford University Press.

Hay, I. (2016) 'On being ethical in geographical research', in N. Clifford, M. Cope, and N. Gillespie (eds) *Key Methods in Geography*, London: Sage, pp. 30–34.

Hay, I. and Israel, M. (2008) 'Private people, secret places: Ethical research in practice', in M. Solem, K. Foote, and J. Monk (eds) *Aspiring Academics*, New York: Prentice-Hall, pp. 165–176.

Healey, R.L. and Ribchester, C. (2016) 'Developing ethical geography students? The impact and effectiveness of a tutorial-based approach', *Journal of Geography in Higher Education*, 40 (2): 302–319.

Herlihy, P.H. (2010) 'Self-appointed gatekeepers attack the American Geographical Society's first Bowman expedition', *Political Geography*, 8 (29): 417–419.

Howitt, R. and Stevens, S. (2016) 'Cross-cultural research: ethics, methods, and relationships', in I. Hay (ed) *Qualitative Research Methods in Human Geography*, Don Mills, Canada: Oxford University Press, pp. 45–75.

Israel, M. (2015) *Research Ethics and Integrity for Social Scientists: Beyond Regulatory Compliance*, 2nd edition, London: Sage.

Israel, M. (2016) 'A history of coercive practices: The abuse of consent in research involving prisoners and prisons in the United States', in M. Adorjan and R. Ricciardelli (eds) *Engaging with Ethics in International Criminological Research*, London: Routledge, pp. 69–86.

Israel, M. (2018) 'Ethical imperialism? Exporting research ethics to the global south', in R. Iphofen and M. Tolich, (eds) *The SAGE Handbook of Qualitative Research Ethics*, London: Sage, pp. 89–102.

Israel, M. and Hay, I. (2006) *Research Ethics for Social Scientists: Between Ethical Conduct and Regulatory Compliance*, London: Sage.

Johnson. J. and Madge, C. (2021) 'Empowering methodologies: feminist and Indigenous approaches', in I. Hay and M. Cope, (eds) *Qualitative Research Methods in Human Geography*, 5th edition, Don Mills, Canada: Oxford University Press, pp. 60–78.

Kearns, R. (1998) 'Interactive ethics: Developing understanding of the social relations of research', *Journal of Geography in Higher Education*, 22 (3): 297–310.

Kindon, S. (2021) 'Empowering approaches: Participatory action research', in I. Hay and M. Cope, (eds) *Qualitative Research Methods in Human Geography*, 5th edition, Don Mills, Canada: Oxford University Press, pp. 309–329.

Kounadi, O. and Leitner, M. (2014) 'Why does geoprivacy matter? The scientific publication of confidential data presented on maps', *Journal of Empirical Research on Human Research Ethics*, 9 (4): 34–45.

Lamberts, R., Grant, W.J., and Martin, A. (2010) 'Public opinion about science', http://cpas.anu.edu.au/files/ANU%20Poll%202010%20Public%20Opinion%20About%20Science.pdf (accessed 27 February 2019).

Lavallée, L.F. and Leslie, L.A. (2016) 'The ethics of university and indigenous research partnerships', in P. Blessinger and B. Cozza (eds) *University Partnerships for International Development* (*Innovations in Higher Education Teaching and Learning, Volume 8*), Emerald Group Publishing Limited, pp. 157–172.

Malički, M., Katavić, V., Marković, D., Marušić, M., and Marušić, A. (2017) 'Perceptions of ethical climate and research pressures in different faculties of a university: Cross-sectional study at the university of split', Science and Engineering Ethics, doi.org/10.1007/s11948-017-9987-y.

Marcus, A.A. (2017) 'Inclusion of personal correspondence' in evolution paper prompts retraction', https://retractionwatch.com/2017/06/30/inclusion-personal-correspondence-evolution-paper-prompts-retraction-new-journal-policy/ (accessed 01 January 2019).

Marcuse, P. (1985) 'Professional ethics and beyond: Values in planning', in M. Wachs (ed) *Ethics in Planning*, New Brunswick: State University of New Jersey Press, pp. 3–24.

Markham, A. and Buchanan, E. (2012) 'Ethical decision-making and internet research', http://aoir.org/reports/ethics2.pdf (accessed 01 January 2019).

Martín, Y., Li, Z., and Cutter, S.L. (2017) 'Leveraging Twitter to gauge evacuation compliance: Spatiotemporal analysis of Hurricane Matthew', *PLOS ONE*, 12 (7). doi.org/10.1371/journal.pone.0181701.

McCoy, A.W. (2006) *A Question of Torture: CIA Interrogation, from the Cold War to the War on Terror.* New York: Henry Holt.

Morris, N. (2015) 'Providing ethical guidance for collaborative research in developing countries', *Research Ethics*, 11 (4): 211–235.

National Commission for the Protection of Human Subjects of Biomedical and Behavioral Research (NCPHSBBR) (1979) 'Belmont Report: Ethical principles and guidelines for the protection of human subjects of research', Department of Health, Education and Welfare, Office of the Secretary, Protection of Human Subjects.

Necker, S. (2014) 'Scientific misbehavior in economics', *Research Policy*, 43 (10): 1747–1759.

Office for Human Research Protections (2001) 'OHRP letter to Hopkins on decision to suspend funding', http://www.baltimoresun.com/bal-ohrpletter-story.html (accessed 02 February 2019).

Olser, R. (2011) '10 reasons why we need social science', *Academy of Social Sciences*, https://campaignforsocialscience.org.uk/news/10-reasons-why-you-need-social-science/ (accessed 28 January 2019).

Oransky, A.I. (2010) 'Authors plan to appeal global ecology and biogeography retraction', https://retractionwatch.com/2010/12/21/authors-plan-to-appeal-global-ecology-and-biogeography-retraction/ (accessed 28 January 2019).

Oransky, A.I. (2014) 'Want to make sure your paper gets published? Just do your own peer review like this researcher did', https://retractionwatch.com/2014/03/20/want-to-make-sure-your-paper-gets-published-just-do-your-own-peer-review-like-this-researcher-did/ (accessed 28 January 2019).

Presidential Commission for the Study of Bioethical Issues (2011) *'Ethically Impossible' STD Research in Guatemala from 1946 to 1948.* Washington DC, https://bioethicsarchive.georgetown.edu/pcsbi/node/654.html (accessed 27 February 2019).

Proctor, J.D. (1998) 'Ethics in geography: giving moral form to the geographical imagination', *Area*, 30 (1): 8–18.

Ramnath, K., Cheaves, S., Buchanan, L., and Borror, K. (2016) 'Incident reports and corrective actions received by OHRP', *IRB: Ethics and Human Research*, 38 (6).

Resnik, D.B. (2015) 'What is ethics in research & why is it important?', https://www.niehs.nih.gov/research/resources/bioethics/whatis/index.cfm (accessed 28 January 2019).

Reverby, S.M. (2009) *Examining Tuskegee: The Infamous Syphilis Study and Its Legacy*, Chapel Hill: The University of North Carolina Press.

Risen, J. (2015) 'American Psychological Association Bolstered C.I.A. Torture Program Report Says', https://www.nytimes.com/2015/05/01/us/report-says-american-psychological-association-collaborated-on-torture-justification.html (accessed 28 January 2019).

Robinson, W. and Unruh, B. (2008) 'The hepatitis experiments at the Willowbrook State School', in E.J. Emanuel, C.C. Grady, R.A. Crouch, R.K. Lie, F.G. Miller, and D.D. Wendler (eds) *The Oxford Textbook of Clinical Research Ethics*, Oxford: Oxford University Press, pp. 80–85.

Robson, E. (2018) 'Ethics committees, journal publication and research with children', *Children's Geographies*, 16 (5): 473–480.

Schrag, Z.M. (2010) *Ethical Imperialism: Institutional Review Boards and the Social Sciences, 1965–2009*, Baltimore: Johns Hopkins University Press.

Sheehan, M., Dunn, M., and Sahan, K. (2018) 'In defence of governance: Ethics review and social research', *Journal of Medical Ethics*, 44 (10): 710.

Smith, D.M. (1997) 'Geography and ethics: A moral turn?', *Progress in Human Geography*, 21 (4): 583–590.

Solomon, J. (2007) *Corporate Governance and Accountability*, 2nd edition, Chichester: Wiley.

Steinberg, P.E. (2010) 'Professional ethics and the politics of geographic knowledge: The Bowman expeditions', *Political Geography*, 29 (8): 413.

Steinberg, P.E., Bryan, J., and Herlihy, P.H. (2011) 'Discussion: Responses to the Bowman expedition editorials', *Political Geography*, 30 (2): 110.

Suspitsyna, T. (2010) 'Accountability in American education as a rhetoric and a technology of governmentality', *Journal of Education Policy*, 25 (5): 567–586.

Taylor, L. (2015) 'No place to hide? The ethics and analytics of tracking mobility using mobile phone data', *Environment and Planning D: Society and Space*, 34 (2): 319–336.

Tobias, J.K., Richmond, C.A.M., and Luginaah, I. (2013) 'Community-based participatory research (CBPR) with indigenous communities: producing respectful and reciprocal research', *Journal of Empirical Research on Human Research Ethics*, 8 (2): 129–140.

Tolich, M. (2010) 'A critique of current practice: Ten foundational guidelines for autoethnographers', *Qualitative Health Research*, 20 (12): 1599–1610.

UK Statistics Authority (2021) 'Ethical considerations in the use of geospatial data for research and statistics', https://uksa.statisticsauthority.gov.uk/publication/ethical-considerations-in-the-use-of-geospatial-data-for-research-and-statistics/ (accessed 19 July 2021).

University of Exeter (2018) 'Ethics', http://geography.exeter.ac.uk/staffarea/ethics/ (accessed 28 January 2019).

Valentine, G. (2005) 'Geography and ethics: Moral geographies? Ethical commitment in research and teaching', *Progress in Human Geography*, 29 (4): 483–487.

van den Bemt, V., Doornbos, J., Meijering, L., Plegt, M., and Theunissen, N. (2018) 'Teaching ethics when working with geocoded data: a novel experiential learning approach', *Journal of Geography in Higher Education*, 42 (2): 293–310.

van den Hoonaard, W.C. (2011) *Seduction of Ethics: Transforming the Social Sciences*, Toronto: University of Toronto Press.

von Benzon, N. and van Blerk, L. (2017) 'Research relationships and responsibilities: 'Doing' research with 'vulnerable' participants: introduction to the special edition', *Social and Cultural Geography*, 18 (7): 895–905.

Voosen, P. (2016) 'The Oaxaca incident', https://www.chronicle.com/article/The-Oaxaca-Incident/236257 (accessed 28 January 2019).

Vujakovic, P. and Bullard, J. (2001) 'The ethics minefield: Issues of responsibility in learning and research', *Journal of Geography in Higher Education*, 25 (2): 275–283.

W3C Interest Group (2021) The Responsible Use of Spatial Data. W3C Interest Group Note 27 May 2021. https://www.w3.org/TR/responsible-use-spatial/ (accessed 19 July 2021).

Walsh, A.C. (1992) 'Ethical matters in pacific island research', *New Zealand Geographer*, 48 (2): 86–86.

Yale University (2018) 'The importance of conducting research ethically', https://assessment-module.yale.edu/human-subjects-protection/importance-conducting-research-ethically (accessed 28 January 2019).

Yanar, Z.M., Fazli, M., Rahman, J., and Farthing, R. (2016) 'Research ethics committees and participatory action research with young people: The politics of voice', *Journal of Empirical Research on Human Research Ethics*, 11 (2): 122–128.

Zhang, J. (2017) 'Research ethics and ethical research: Some observations from the global south', *Journal of Geography in Higher Education*, 41 (1): 147–154.

Žižek, S. (2000) *Enjoy Your Symptom!: Jacques Lacan in Hollywood and Out*, New York: Routledge.

3 Research ethics in human and physical geography

Ethical literacy, the ethics of intervention, and the limits of self-regulation

Susann Schäfer

Introduction

Research ethics are increasingly important for scientists who want to act with integrity in their role as researchers (Israel 2015). Research ethics can be distinguished from scientific ethics: While research ethics focus on the relationship between researchers and participants in empirical research, the ethics of science deals with the rules of good scientific practice (Döring and Bortz 2015: 122). The goal of ethical research is the protection of the subjects' dignity and welfare: Neither humans nor animals should be unethically abused or harmed for scientific purposes in empirical studies (ibid.).

In a large survey among scientists from different academic fields in the United States, Martinson and his colleagues discovered that a large number of (especially young) researchers face ethical dilemmas and do not behave according to ethical scientific standards (Martinson, Anderson and Vries 2005). Thus, researchers might fail to present data that contradicts their own previous research or change the design, methodology, or results of a study in response to pressure from the funding source. However, the increased importance of research ethics is not only driven by a heightened sensitivity towards ethical questions within scientific communities, but is also due to new technological developments (opening up new research possibilities), the globalisation of research (i.e. the fact that research projects are becoming more and more internationally oriented), the emancipation of societies towards science (research is now critically questioned with regard to unjustified claims to authority and power), and finally science scandals (ensuring that existing measures to prevent unethical behaviour by researchers are reviewed and optimised) (Döring and Bortz 2015: 122).

As a result, many scientific communities have intensified discussions on research ethics (Kara 2018). These debates are characterised by the emphasis on applied, and not theory-driven, research ethics giving credit to the fact that the existence of clear ethical principles is not enough to guarantee ethical research practices due to the lack of implementation of ethical guidelines and the emergence of new ethical problems in the implementation of ethical standards (Russell 2013: 113). These thoughts

DOI: 10.4324/9780429507366-3

are also becoming increasingly relevant to geography. Due to the strong fragmentation of this discipline, the status-quo of research ethics is very heterogeneous.

When geography was established as a scientific discipline at the end of the 19th century, geographers were equally concerned with both the physical and the human geographical dimensions in the scholarly description and analysis of regional-specific phenomena. Over time, geography soon developed into two fundamental subfields – the physical and the human sphere – with specific theories, methodologies, and methods (Livingstone 2009). Even within these subfields today there are manifold theoretical and methodological ramifications: Whereas physical geography addresses research questions in climatology, geomorphology, soil geography, and many other areas, human geography has developed into further nuanced branches, such as social, political, and economic geography. However, this heterogeneity should not hide the fact that for several years now there has been an increased cooperation between physical and human geography – the 'third convergence' (Simon and Graybill 2012) – due to phenomena that require holistic research perspectives, such as climate change, the provision of energy and resources, and global food systems (Baerwald 2010). This branch of geographical research seeks to analyse both dynamics in the physical sphere as well as processes and governance of the related social and political systems in order to derive sustainable policy actions (Goudie 2017). It is in these interdisciplinary research contexts in particular, but not exclusively in them that the issue of research ethics becomes a critical one, addressing the question of how and for which purpose research is undertaken. Assuming that physical and human geographers have different perspectives on these questions, research ethics can emerge as a field of tension in such projects.

Against the background of these disciplinary developments, this chapter addresses the question of research ethics with a focus on human and physical geography. In particular, the chapter deals with the differences and commonalities of research ethics between human and physical geography. The chapter is structured as follows: The following section describes the most important findings of a survey on research ethics conducted among geographers in Germany, Switzerland, and Austria. The subsequent section discusses the underlying causes and explanations of the survey results. The next section outlines ethical issues raised in research projects at the nature-society interface, specifically on climate change adaptation and ecosystem management. The ethical issues raised are exemplary for research cooperation between human and physical geographers but are not limited to these kinds of geographical studies. The subsequent section goes on to look at how ethical standards and practices in physical-human geographical research projects can be realised. Finally, the chapter concludes with a summary of the major findings and indicates the direction of future research.

Research ethics – a topic of equal importance for human and physical geographers?

In 2017, geographers working at universities and research institutes in Germany, Austria, and Switzerland were invited to participate in an online survey[1] about their attitudes towards and experiences with ethics in their empirical research. The survey confirmed the assumption that there are indeed commonalities and differences between human and physical geographers in terms of (a) individual experiences with research ethics, (b) research phases where challenges regarding research ethics arose for individual researchers, and (c) the general assessment of research ethics for the geographical subdiscipline (e.g. economic geography, soil science). The response rate of 22% of physical geographers and 78% of human geographers already indicates that the topic of the survey – research ethics – received greater attention in the field of human geography compared to physical geography. This first impression was reinforced on a closer examination of some of the selected results.

When it comes to individual experiences with ethical challenges during empirical research, 70% of human geographers and only 47% of physical geographers said that they had already faced ethical problems in their own research. Around 8% of human geographers and 26% of physical geographers reported that they were not able to answer this question. Both figures indicate that, first, human geographers are more aware of ethical issues – either because they are confronted with more ethical challenges per se or because they have developed a greater sensitivity towards ethical questions or 'ethical literacy' during their scientific socialisation. Second, the relatively high proportion of physical geographers who felt unable to answer the question suggests that research ethics were not considered to be essential in research practices and overall they could therefore have a greater ignorance of research ethics.

Those geographers who had experienced ethical challenges were asked further on in the questionnaire when these problems arose in the research process. The researchers behind the online questionnaire specified the following research phases: Definition of the research subject, development of

---

1   The survey was designed and carried out by Sebastian Henn, professor of economic geography at the Friedrich Schiller University Jena, Germany, Judith Miggelbrink, professor of human geography at the Technical University Dresden, Germany, and Susann Schäfer, post-doctoral researcher in economic geography at the Friedrich Schiller University Jena, Germany.

The survey took place between 14 September and 31 October 2017. The link to the online survey was send to all members of the Association for Geography at German speaking universities and research facilities (Germany, Austria, Switzerland). In total, 179 geographers participated in this study. Among those who answered the questionnaire were 22% physical geographers and 78% human geographers. The data set with 179 cases was analysed with the computer program SPSS.

the research design, application for funding, attempts to access the research field, the process of data collection and analysis, publication and presentation of research results, and finally the development of policy recommendations.

Starting with the definition of the research subject, roughly the same amount of human and physical geographers reported having faced ethical questions. Around 52% of human and physical geographers experienced an ethical issue regarding the selection of research subjects. When it comes to research ethics during the development of a research design, there is a considerable gap between human geographers (40%) and physical geographers (7%). Obviously, research ethics for physical geographers plays a subordinate role in this phase, while almost every second human geographer deals with ethical questions. In terms of application for funding, every fifth human geographer (22%) has experienced ethical issues, whereas only every tenth physical geographer reported facing ethical questions. Ethical issues related to access to the field are also more widespread among human geographers (57%), although a major share of physical geographers (45%) experience similar issues here, too. In terms of research ethics in data collection and analysis, the share of affected geographers is high for both groups (human geographers 60%, physical geographers 49%). As for the publication and presentation of research results (human geographers 50%, physical geographers 45%) as well as the development of policy recommendations (human geographers 22%, physical geographers 21%), there is no considerable gap between human and physical geographers. These descriptive statistics indicate that ethical problems arise during data collection and analysis in particular.

Whereas these results refer to the individual experiences of the researcher, the respondents were also asked to assess the role of research ethics in the proposed research phases for their entire subcommunity of human or physical geographers. One important finding is that human geographers consider research ethics to be more relevant than physical geographers for the respective subdiscipline. Almost 50% of human geographers think that research ethics play a major or very major role in the selection of research subjects – compared to 23% of physical geographers. Another interesting result is that the assessment of whether research ethics play a very major or major role in the respective research phase has a clear tendency among human geographers. Compared to them, the assessment of how relevant research ethics of physical geography does not follow a trend. Whereas human geographers clearly label research ethics as an important issue in terms of selecting the research subject, the role that ethics play here in physical geography is heterogeneous and does not reflect a clear tendency. When it comes to the development of research concepts, the evaluation of the role of research ethics between human and physical geography is the opposite. A large share of physical geographers (54%) says that research ethics play little or no role in the design of research in the entire community of physical geographers. For human geographers, only 10% say that research ethics have little or no meaning in this research phase. With regard to the importance of research ethics in the application for the funding process, there are no clear tendencies within the sub-areas of geography (physical or human), but also between physical and human geography. At most, there is a slight tendency for human geography to rate its significance somewhat higher. However, these figures are not particularly strong. When it comes to accessing the research field there are again differences: In human geography, this part of the research is classified as relevant. For around 60% of the respondents, ethical questions play a major or very major role with regard to assessing the research fields. In physical geography, the ethical dimension of field access is assessed very diversely. There is no tendency as to whether the community rates ethics in field access as high or low. A similar picture can be seen in data collection and analysis. While human geographers consider ethical questions important or very important for their community (59%), there is no trend in physical geography. An interesting picture emerges from the publication of research results: For human geographers, ethical questions tend to be relevant here (49% say that they are important or very important, whereas only 8% say that ethical questions play little or no role). The group of physical geographers, on the other hand, is divided. Either physical geographers find ethical questions important or very important (36%) or to have little or no importance (31%). The role of ethics in the development of recommendations for action is similarly important for physical and human geographers. Forty-two percent of human geographers and 51% of physical geographers consider ethical considerations to be important or very important in the development of recommendations for action.

The results of this survey indicate that research ethics have a different status in human and physical geography. In general, research ethics are more established in human geography compared to physical geography. It is striking that in the evaluation of research ethics among physical geographers, trends are often not discernible in the respective research phases, suggesting that in this subgroup there is no consensus on this aspect of research practices. There are similarities and differences between physical and human geographers with regard to the research phase when ethical questions are assessed as relevant. Both physical and human geographers perceive the selection of research subjects and data collection/analysis to be an ethical decision, represented by a considerable share of both physical and human geographers. However, human geographers generally assign greater importance to research ethics in the other research phases compared to physical geographers.

Explaining the gap of research ethics between human and physical geography

When we discuss the commonalities and differences of research ethics among human and physical geographers, we need to acknowledge that they are both fragmented subfields in the discipline of geography that are characterised by specific discourses, methods, and research practices. Although research ethics are addressed in introductory textbooks for both physical and human geography students (Boyd et al. 2008; Gomez and Jones 2010; Hay 2016; Montello and Sutton 2013; Smith 2010), research ethics have a much longer tradition in human geography compared to physical geography. Human geographers, especially those working with qualitative methods, have debated the ethics of research practices for a long time (Barnett 2012; Tadaki, Slaymaker and Martin 2017). In social and cultural geography, constructivism has emerged as the scientific paradigm after the 'cultural turn' in the 1990s (Mitchell 2000) acknowledging the situatedness of empirical research, the politics of knowledge production, and the co-existence of multiple truths. This paradigm shift has led to a different approach to research topics and subjects.

Among the research topics that raised critical ethical concerns, for example, was the introduction of GIS technology, initially a military tool, in human geographical research. In the 1990s, many human geographers opposed the use of GIS technology for research purposes based on epistemological issues and ethical dilemmas (Schuurman 2000). When GIS became an indispensable tool in human geography, critics proposed an 'ethics of GIS' (Crampton 1995). However, physical geographers were mostly not concerned with the use of such technology in their research.

The changed attitude towards research subjects was reflected in a more responsible and transparent approach to the whole research process, which regarded research subjects not only as a means to an end but as carriers of knowledge relevant to the research. This in turn has led to the emergence of institutional standards that now have to be implemented by researchers in many research institutions (e.g. approval through ethical review, an informed consent of the research participants). This institutionalisation process developed parallel to the standardisation of ethical guidelines in the life sciences and other social sciences (Kapp 2006). In physical geography, compared to human geography, it can be argued that ethical reflections on research practices played a very minor role during the 1990s (Proctor 1998: 13), partly attributed to the positivist paradigm which is widely accepted in the natural sciences and hence in physical geography. Positivist thinking underlines the existence of one truth and scientific knowledge production as a linear process marked by verification and falsification statements (Kitchin 2015). Positivist thinking does not per se reject ethical considerations. However, physical geography did not experience a 'cultural turn' with its strong impact on research practices in human geography. It is only more recently that research ethics have become a debated topic in physical geography (Blue 2018; Ford et al. 2016; Powell 2002), although they are still far from becoming a mainstream issue in the subdiscipline. The major reason for the low significance of research ethics is that they seem to 'require skills in philosophical, historical and sociological analyses that can seem peripheral to the task at hand and especially unrelated to research questions, for example, in the biophysical sciences, in theoretical physics or indeed the life cycle of the European eel' (Russell 2013: 114). However, the lack of ethical consideration comes at a cost: The ignorance of research ethics during the entire research process leads to a lack of necessary contextual knowledge required to develop meaningful policy recommendations. Hoegh-Guldberg et al., for example, argue that management interventions after conclusion of the data analysis tend to be too general and do not acknowledge cultural, political and/or economic specifics of the governance system or society (Hoegh-Guldberg et al. 2007). As a result, policy recommendations are detached from social reality and thus have no effect, as they do not take into account the respective framework conditions.

The survey presented in the previous section indicated that ethical questions in human and physical geography often emerge at different stages of the research process. In human geography, ethical considerations are made from the beginning of the research design process, as a process is put in place to ensure research participants are able to give informed consent to surveys, interviews, participant observations, and other social science methods (Watts 2011; Winchester 1996). Concerns around gaining access to research participants, their involvement in the research, and the implications for them when participating in the research are critical points that require individual (from research subjects) as well as perhaps also institutional (from research institutions) consent before the researcher can undertake his/her fieldwork (von Benzon and van Blerk 2017). According to the literature, physical geographers are, by comparison, predominantly faced with the 'ethics of intervention' (Robbins and Moore 2013), at the end of research projects, with regard to the implications of results for the governance of natural and social systems (Blue 2018). This is far from a trivial task, given that results often include probabilities and uncertainties that do not allow policy actions to be derived directly. The best example is research on climate change where scientists have developed different scenarios that can have diverse climate change impacts. Policy recommendations therefore also differ widely, depending on the scenarios developed (Moss et al. 2010). The statement that physical geographers are mainly concerned with the 'ethics of intervention' contradicts to a certain extent the results of the survey, which clearly show that ethical considerations are also made by physical geographers before they are concerned with the development of policy recommendations. However, these ethical challenges are hardly addressed in the scientific debate. Kershaw et al. (2014) indeed point out that there are potential ethical issues that apply to physical geographical research at earlier phases, such as access to land with specific indigenous property/use rights relevant for data collection (Kershaw, Castleden and Laroque 2014). Often, data collection and analysis of natural system elements (e.g. water or soil) are not considered to infringe upon the personal rights of individuals (except for property rights). Nevertheless, this kind of data collection and analysis poses ethical questions regarding power relations between researchers, regional representatives, and/or indigenous populations.

The differences between research ethics are potentially a source of conflict in projects belonging to the 'third convergence' – research that stretches both physical and human geographical questions. In those projects, geographers often experience theoretical tensions, even sometimes incompatibilities, and methodological challenges (Lave 2015; Massey 1999), that are not only linked to diverging approaches to ethics but to different scientific paradigms and research practices (Harrison et al. 2004). In the following section, one research example is used to highlight exemplary ethical problems in interdisciplinary projects and how these were overcome by the researchers conducting those projects. Although the examples are specific with regard to the study region and the methods applied, the ethical problems experienced can be expected to emerge in interdisciplinary geography projects.

Research ethics in projects from human and physical geography

Examples of community-based adaptation and ecosystem management

In order to take a closer look at the precise ethical issues involved in projects at the human-nature interface and how these issues can be resolved, two examples will be presented that focus on climate change adaptation and ecosystem management, which are research fields where many interdisciplinary studies have been conducted over the past 10 years (Webber 2016). The project example published by Ford et al. (2016) is a community-based climate change adaptation project in the Canadian Arctic. Community-based adaptation is a participatory method in climate change research that enables communities to cope with climate change impacts (van Aalst, Cannon and Burton 2008). The general idea is that local communities and researchers develop coping mechanisms for the changing environment and implement these mechanisms together. In this article (Ford et al. 2016), 23 scientists who conducted community-based adaptation projects in Northern Canada described the ethical challenges during their research. During a two-day workshop, they discussed the ethical aspects of adaptation research that was conducted in cooperation with Inuit communities in Canada, and eventually published the results of their debate on research ethics. Ford et al. (2016) summarise three ethical problems that are likely to emerge in comparable circumstances in research projects from the disciplines of human and physical geography.

1 *Scientific vs. indigenous knowledge*: The scientific discourse on climate change adaptation (Adger et al. 2009) has the potential to perpetuate the legitimation of research interventions in the governance of local communities while creating unequal power relations between researchers coming from the outside and indigenous communities. The researchers

in Ford et al.'s (2016) project experienced moments during which their scientific knowledge on local climate change impacts contradicted the indigenous knowledge of the community. Crucially, they understood that their position in the research project could lead to the devaluation of indigenous knowledge and thus also to a devaluation of the research participants involved. As a consequence of this discrepancy, the researchers understood that they had to proceed sensitively when evaluating and transmitting scientific knowledge in order not to violate the relationship with the indigenous group.

2 *The multiple roles of researchers*: By applying this kind of participatory method it led to close relations with the local community. The geography researchers reported having multifaceted roles in community-based adaptation projects, for instance as educators, communicators, community workers, promoters, etc. However, 'such demands and responsibilities can also compromise the integrity and quality of the research component of projects. Several researchers reported that they were being frequently asked to provide additional services for the indigenous communities, such as educational resources and training outside of the research parameters, and requests to look for public funds' (Ford et al. 2016: 183). However, during the project, it became an ethical dilemma for some researchers to receive requests from the community that were partly in conflict with the allotted time and funding.

3 *Sustainability of the research*: The question as to whether the community-based adaptation project could achieve a positive change during the period of funding raised another ethical concern for the researchers (Ford et al. 2016: 179). Some of the researchers in the project saw the risk of 'over-promising' results and outcomes. Because its funding was short-term, some researchers believed that the project could present a sustainable result during the project period. They faced the following dilemma: On the one hand, the researchers needed to develop an evidence base on adaptation, on the other hand, the participatory process used to create this evidence base could have generated significant expectations for positive change, which, if not followed by visible changes, could have reduced local interest in and valuation of the research or, worse, become maladaptive (Ford et al. 2016: 180).

The lessons learned for the researchers in the community adaptation project were that they became aware of the complexities of conducting participant research with indigenous populations. Critical aspects arose from the tension between the limited project duration and the time frame of indigenous groups, between expectations and roles, and questions how sometimes vague scientific results can be transformed into precise policy actions that would take longer than the project period. The authors concluded that the consideration of these ethical questions was time consuming, but allowed a more sustainable impact on the climate change adaptation process. The focus on how researchers deal ethically with the results of their projects in the context of decisions will be the subject of the second example.

Using the example of a river, Blue argues that researchers 'wrestle with the ethics of intervention' (Blue 2018: 462) because conceptions of 'healthy' or desirable states of ecosystems are not pre-defined but socially constructed – adding a fourth dimension to the list of ethical challenges:

4   *Socially constructed management norms*: Although there is a societal con-sensus that improved ecosystem conditions can be considered as a valu-able aim, precise conceptions of how to measure and reach that goal are less clear. Thus, geographers seeking to develop policy recommendations are challenged with the ethics of intervention. Blue convincingly outlines how environmental ideals have transformed over time and how there has been a shift from holistic concepts to measurable analysis and evaluation of ecosystem management. For interdisciplinary projects, this means that the inherent ethical consideration for geographers goes beyond the ques-tion of how results from a physical geographical analysis translate into rec-ommended actions, especially when the results are linked to probabilities and imply a degree of uncertainty, for example concerning future climate projections (Hulme 2012). In addition to this consideration, researchers need to acknowledge that there might be different desirable states of eco-system management potentially leading to different types of policy inter-vention. To make meaningful decisions, researchers need to acknowledge the social, economic, and political contexts of ecosystem management.

The solution that both author teams suggested was the collective dialogue on issues that at least one research or participant partner felt uncomfortable with. Sharing one's ethical beliefs and assumptions can be a fruitful opportunity for mutual learning and understanding in interdisciplinary projects. In the worst case, research ethics will remain the responsibility of the single researcher and a joint debate on the specific challenges of research ethics will be missing.

Ethics of intervention: ecosystem management and policy recommendations

The ethics of intervention and the other ethical questions presented in the previous section point to different challenges in the research process. Whereas the adaptation project debated different expectations and roles of the subjects involved and how to balance these diverging expectations, the ethics of intervention point to a more fundamental issue in interdisciplinary projects. The leaders of the adaptation project argued that continual dia-logue and transparency between researchers and communities, beginning at project inception, made all involved parties aware of the direction of the work and helped them to minimise misunderstandings (Ford et al. 2016). Moreover, the responsible researchers underlined that academic, Western knowledge was not privileged and that they tried to balance different types of knowledge. Communication between all research partners was key to solving critical ethical aspects. Ford et al. summarise their communication style as follows: 'Key traits necessary for work of this nature include deep listening, patience, openness to multiple ways of knowing, willingness to accept and respond to criticism, flexibility, self-reflection, an ability to communicate and facilitate, a willingness to learn, a desire to develop and maintain strong and lasting relationships, and a sense of humor' (2016: 186).

Although ethical practices within research projects pose a challenge, the 'ethics of intervention' seem to be even more complex given that they include the negotiation of contested and sometimes contradictory visions of the desirable states of environmental ecosystems. This is similar to Russell's argument that ethical reasoning can support researchers in this negotiation by making explicit the unconsciously accepted or critiqued value judgements (Russell 2013). Developing visions of desirable states of such ecosystems includes the consideration of different stakeholder groups and their needs and wishes (Salisu Barau, Stringer and Adamu 2016). In other words, the intervention needs to be based on understanding both the social and the physical processes that shape ecosystems. This includes the understanding that quantitative approaches, such as physical-geographical models, do not have a 'monopoly on what matters' (Blue 2018: 470). Through considering the social and physical processes it will be possible to produce a clearer picture of which interventions are possible and desirable (Blue 2018: 471). Understanding these processes requires interdisciplinary cooperation and may bring physical and human geographers closer together in shaping desirable futures of ecological systems. Critical attention to relations of social power and processes with deep knowledge of a particular field of physical geography in the service of social and environmental transformation is a beneficial perspective at the interface of human and physical geography (Goudie 2017: 20).

While the ethical questions in the community-based adaptation project addressed the way in which researchers and subjects interacted, the ethical intervention of ecological systems goes beyond this internal perspective and requires a more fundamental change in thinking and cooperation.

Strengthening research ethics – is self-regulation enough?

The previous two examples show that reflections during the research process have led scholars to discuss and solve ethical questions on the basis of self-regulation. The scholars of the two examples stressed this kind of individual regulation and did not emphasise the meaning of external regulation, for example, ethical reviews through university committees, to ensure compliance with ethical standards.

Back in the 1980s, Mitchell and Draper proposed the following four approaches to resolve ethical conflicts in geographical research (Mitchell and Draper 1983): individual self-regulation, disciplinary responses, institutional controls, and external controls. Individual self-regulation is based

Table 3.1 Differences between self-regulation and external regulation

Category	Self-regulation	External regulation
Responsibility	Single researcher	Ethics committee or other
Tools and measures	Reflection, thinking, seeking help optional	Codes of conduct, ethics assessment
Norms	Personal values and experiences	Institutional norms and regulations
Time	Any time during research	Usually practiced at the project start
Control and intervention	No control, nor external intervention	Control and intervention through approval or rejection of research plan

on the reliance upon conscientious researchers to ensure that research is conducted in an ethical way (see Table 3.1). Disciplinary responses address the issue of formal codes of ethics and the establishment of ethics committees, such as the statement of the American Association of Geographers on 'Ethical Behaviour during Field Research' (Association of American Geographers 2018). Institutional controls shift the review of ethical conduct to research grant committees and funding agencies. There are also external controls related to government guidelines and regulations, such as the data protection ordinance. In general, research ethics can be regulated between the two extremes: self-regulation and external regulation.

In general, approaches to strengthen research ethics in geography and other social sciences have developed from principle-based conceptions of ethical conduct to more flexible forms of ethical reflection (Russell 2013: 113). However, external control plays an important role in many university systems by controlling and limiting research topics. In addition, dealing with external control requires resources from the researcher. Not only for this reason but also for a number of other reasons, research ethics boards for the compliance with ethical standards in research have come under criticism. According to Haggerty (2004: 410), 'there is often a distinct but unquestioned rupture between following the rules and conducting ethical research'. Haggerty justifies his statement with the fact that rigid external controls lead to researchers using all means to force the approval of their projects. Sometimes researchers behave unethically in order to obtain this approval, which in turn calls into question the meaning of a process. Another point of criticism is that ethical ambiguity and tensions cannot always be solved through external regulation and controls, because, firstly, they are not known at the onset of the research or, secondly, they are not covered by the content of ethical standards. Thus, the diversity of potentially arising ethical issues in research contexts cannot be sufficiently covered by institutional forms and regulations. For geographical research, for example, the American Association of Geographers (AAG) proposes that field research should be conducted in ways that minimise long-term impacts (Association of American Geographers 2018). The statement suggests 'efficient sampling procedures' and encourages researchers to respect and preserve the natural environment as much as possible (Association of American Geographers 2018). These specifications are certainly correct in their statements, but it is quite possible that in concrete research contexts ethical questions may arise for which there is no formal solution.

Against the background of this criticism, many geographers think that the ongoing ethical debates in human geography are not supported but rather threatened by such review systems (Dyer and Demeritt 2009; Hay 1998). There is a strong trend towards individual or collective self-regulation of research ethics carried by the conviction that ethical relationships between researchers and research subjects are characterised by an ongoing interrogation of the types of responsibilities that researchers might owe to others (Bauman 1993). It is argued that those relationships cannot be reduced to the simple exercise of following a set of ethical standards and it becomes apparent that the application of many of the existing rules bears little relationship to ethical conduct whatsoever (ibid.). Self-regulation, however, requires certain commitment and skills besides the researcher: Not only must the researcher be willing to consider research-ethical issues and sometimes draw the negative consequences (e.g. termination of the research in case of ethical conflicts), but she must also be able to 'educate' herself through literature studies or exchanges with colleagues. The reason is that 'ethical literacy' in research is a skill that develops in the course of dealing with ethical problems in research.

In light of the study presented at the beginning of the chapters, it seems questionable whether the path of self-regulation is a viable solution for the entire discipline of geography. It could be argued that self-regulation is the best strategy for researchers who have reached a certain level of 'ethical literacy' and who do not have to accept substantial career disadvantages if they determine or adapt research projects due to ethical problems. Especially for young researchers, the discontinuation of research for ethical but also other reasons can considerably restrict their scientific career. It would therefore make sense not to leave it up to individuals to comply with research ethics but to create an institutional framework in which resources are made available that young researchers can draw from when they are confronted with ethical problems. For example, universities or funding agencies could provide additional funding for those PhD students or postdocs who are faced with the decision whether to breach ethical standards in their research or terminate the research process.

Another aspect of the considerations concerns the level of 'ethical literacy' among researchers. Although it can be assumed that research experience and 'ethical literacy' are correlated, it can nevertheless be assumed on the basis of the data that even among established scientists many have little experience with the self-regulation of ethical issues. This would also be an argument for not relying solely on the individual researcher, but for adopting a collaborative approach. There is a need for a change in practices to discuss ethical questions more openly and in a transparent way in all kinds of informal research meetings and public workshops and conferences expanding 'a culture that prizes such skills and regards them as integral to the excellence of the research process itself' (Hogan 2013: 5).

Conclusion

This chapter sheds some light on a specificity of geography in the context of research ethics: the diversity of the discipline as a human and physical geographical field. Given the different theoretical and methodological foundations in each field, research ethics are evaluated differently among human and physical geographers: While ethical issues are often considered from the onset in human geographical research, in physical geography, they most often gain particular relevance when it comes to the implications of research results. Although these differences can lead to tensions in interdisciplinary projects, they also demonstrate a potential field of synergy and mutual learning.

The project examples of geographical research at the interface show that there are potential ethical tensions in the following areas: Scientific vs. indigenous knowledge, the roles of researchers, the sustainability of research, and multiple potential futures of an ecosystem. According to the authors of these projects, ethical concerns can be solved among project partners if there is an awareness of and openness towards ethical reflection. Alternatively, projects may need to be designed differently and should include new partners to ensure that ethical considerations are not just paid lip service. The only condition for ethical reflection and critical thinking is that the academic community needs to create 'a culture that prizes such skills and regards them as integral to the excellence of the research process itself' (Hogan 2013: 5). This argument propagates the self-regulation of researchers in ethical questions, which, however, are currently regulated in many university systems by institutional processes such as an ethics review. These processes are seen as critical in science, as they do not guarantee compliance with ethical standards. Therefore, it remains to be discussed how effective self-regulation really is and which soft instruments are possible for practices to change in science. To answer these questions, research on research is needed – not only on how prevalent ethical literacy is among researchers but also on how it can be developed and what the obstacles to ethical research are.

Acknowledgements

The author would like to thank, first, Kathrin Hörschelmann and Emma Roe for their critical feedback and suggestions on an earlier draft of the chapter, and second, Judith Miggelbrink and Sebastian Henn for their cooperation in the study on research ethics among German-speaking geographers in Germany, Switzerland, and Austria.

References

Adger, W.N., Dessai, S., Goulden, M., Hulme, M., Lorenzoni, I., Nelson, D.R., Naess, L.O., Wolf, J., and Wreford, A. (2009) 'Are there social limits to adaptation to climate change?', *Climatic Change*, 93 (3–4): 335–354. doi: 10.1007/s10584-008-9520-z.

Association of American Geographers (2018) 'Statement on professional ethics', http://www.aag.org/cs/about_aag/governance/statement_of_professional_ethics (accessed 12 September 2019).

Baerwald, T. (2010) 'Prospects for geography as an interdisciplinary discipline', *Annals of the Association of American Geographers*, 100 (3): 493–501. doi: 10.1080/00045608.2010.485443.

Barnett, C. (2012) 'Geography and ethics', *Progress in Human Geography*, 36 (3): 379–388. doi: 10.1177/0309132510397463.

Bauman, Z. (1993) *Postmodern ethics*. Oxford: Blackwell.

Blue, B. (2018) 'What's wrong with healthy rivers? Promise and practice in the search for a guiding ideal for freshwater management', *Progress in Physical Geography*, 42 (4): 462–477. doi: 10.1177/0309133318783148.

Boyd, W.E. et al. (2008) "None of us sets out to hurt people': The ethical geographer and geography curricula in higher education1', *Journal of Geography in Higher Education*, 32 (1): pp. 37–50. doi: 10.1080/03098260701731462.

Crampton, J. (1995) 'The ethics of GIS', *Cartography and Geographic Information Systems*, 22 (1): 84–89. doi: 10.1559/152304095782540546.

Döring, N. and Bortz, J. (2015) *Forschungsmethoden und Evaluation in den Sozial- und Humanwissenschaften (Research* methods and evaluation in the social and human sciences). Berlin and Heidelberg: Springer.

Dyer, S. and Demeritt, D. (2009) 'Un-ethical review? Why it is wrong to apply the medical model of research governance to human geography', *Progress in Human Geography*, 33 (1): 46–64. doi: 10.1177/0309132508090475

Ford, J.D. et al. (2016) 'Community-based adaptation research in the Canadian Arctic', *Wiley Interdisciplinary Reviews. Climate Change*, 7 (2): 175–191. doi: 10.1002/wcc.376

Gomez, B. and Jones, J.P. (eds.) (2010) *Research methods in geography: A critical introduction*. Chichester: Wiley-Blackwell (Critical introductions to geography).

Goudie, A.S. (2017) 'The integration of human and physical geography revisited', *The Canadian Geographer/Le Géographe Canadien*, 61 (1): 19–27. doi: 10.1111/cag.12315

Haggerty, K. (2004) 'Ethics creep: Governing social science research in the name of ethics', *Qualitative Sociology*, 27 (4): 391–414.

Harrison, S., Massey, D., Richards, K., Magilligan, F.J., Thrift, N., and Bender, B. (2004) 'Thinking across the divide: Perspectives on the conversations between physical and human geography', *Area*, 36 (4): 435–442. doi: 10.1111/j.0004-0894.2004.00243.x.

Hay, I. (1998) 'Making moral imaginations. Research ethics, pedagogy, and professional human geography', *Philosophy & Geography*, 1 (1): 55–75. doi: 10.1080/13668799808573632.

Hay, I. (2016) 'On being ethical in geographical research', in N. Clifford, M. Cope, T. Gillespie and S. French (eds) *Key methods in geography*, Los Angeles: Sage, pp. 30–43.

Hoegh-Guldberg, O., Mumby, P.J., Hooten, A.J., Steneck, R.S., Greenfield, P., Gomez, E., Harvell, C.D., Sale, P.F., Edwards, A.J., Caldeira, K., and Knowlton, N. (2007) 'Coral reefs under rapid climate change and ocean acidification', *Science*, 318 (5857): 1737–1742. doi: 10.1126/science.1152509.

Hogan, L. (2013) 'Developing ethics as a core competency: Integrity in scientific research', in C. Russell, L. Hogan and M. Junker-Kenny (eds) *Ethics for graduate researchers: A cross-disciplinary approach*. Amsterdam: Elsevier, pp. 1–6.

Hulme, M. (2012) 'Climate change', *Progress in Physical Geography*, 36 (5): 694–705. doi: 10.1177/0309133312456414.

Israel, M. (2015) *Research ethics and integrity for social scientists: Beyond regulatory compliance*. 2nd edn. London: SAGE.

Kapp, M.B. (2006) 'Ethical and legal issues in research involving human subjects: do you want a piece of me?' *Journal of Clinical Pathology*, 59 (4): 335–339. doi: 10.1136/jcp.2005.030957.

Kara, H. (2018) *Research Ethics in the Real World*. Bristol: Policy Press.

Kershaw, G.G.L., Castleden, H., and Laroque, C.P. (2014) 'An argument for ethical physical geography research on Indigenous landscapes in Canada', *The Canadian Geographer/Le Géographe canadien*, 58 (4): 393–399. doi: 10.1111/cag.12092.

Kitchin, R. (2015) 'Positivist geography', in S. Aitken and G. Valentine (eds) *Approaches in human geography*. Newbury Park: SAGE, pp. 23–34.

Lave, R. (2015) 'Exploring the proper relation between physical and human geography', *Progress in Physical Geography*, 39 (5): 687–690. doi: 10.1177/0309133315595727.

Livingstone, D. (2009) 'History of geography', in D. Gregory, R. Johnston, G. Pratt, M. Watts and S. Whatmore (eds) *The dictionary of human geography*. Oxford: Wiley-Blackwell, pp. 295–299.

Martinson, B.C., Anderson, M.S., and Vries, R. de (2005) 'Scientists behaving badly', *Nature*, 435 (7043): 737–738. doi: 10.1038/435737a.

Massey, D. (1999) 'Space-time, 'science' and the relationship between physical geography and human geography', *Transactions of the Institute of British Geographers*, 24 (3): 261–276. doi: 10.1111/j.0020-2754.1999.00261.x.

Mitchell, B. and Draper, D. (1983) 'Ethics in geographical research', *The Professional Geographer*, 35 (1): 9–17. doi: 10.1111/j.0033-0124.1983.00009.x.

Mitchell, D. (2000) *Cultural geography: A critical introduction*. Oxford: Blackwell.

Montello, D.R. and Sutton, P.C. (2013) *An introduction to scientific research methods in geography & environmental studies*. 2nd edn. Los Angeles: SAGE.

Moss, R.H. et al. (2010) 'The next generation of scenarios for climate change research and assessment', *Nature*, 463 (7282): 747–756. doi: 10.1038/nature08823.

Powell, R.C. (2002) 'The Sirens' voices? Field practices and dialogue in geography', *Area*, 34 (3): 261–272. doi: 10.1111/1475-4762.00080.

Proctor, J.D. (1998) 'Ethics in geography: Giving moral form to the geographical imagination', *Area*, 30 (1), pp. 8–18.

Robbins, P. and Moore, S.A. (2013) 'Ecological anxiety disorder: diagnosing the politics of the Anthropocene', *Cultural Geographies*, 20 (1): 3–19. doi: 10.1177/1474474012469887.

Russell, C. (2013) 'Contextualising ethical principles in research practice in different disciplines', in C. Russell, L. Hogan and M. Junker-Kenny (eds) *Ethics for graduate researchers: A cross-disciplinary approach*. Amsterdam: Elsevier, pp. 113–119.

Salisu Barau, A., Stringer, L.C., and Adamu, A.U. (2016) 'Environmental ethics and future oriented transformation to sustainability in sub-Saharan Africa', *Journal of Cleaner Production*, 135: 1539–1547. doi: 10.1016/j.jclepro.2016.03.053.

Schuurman, N. (2000) 'Trouble in the heartland: GIS and its critics in the 1990s', *Progress in Human Geography*, 24 (4): 569–590. doi: 10.1191/030913200100189111.

Simon, G. and Graybill, J. (2012): 'Geography in interdisciplinarity: Towards a third conversation', *Geoforum*, 41 (3): 356–363. doi: 10.1016/j.geoforum.2009.11.012.

Smith, D. (2010) 'The politics and ethics of research', in B. Gomez and J.P. Jones (eds) *Research methods in geography: A critical introduction (Critical introductions to geography)*. Chichester: Wiley-Blackwell, pp. 411–424.

Tadaki, M., Slaymaker, O., and Martin, Y. (2017) 'Changing priorities in physical geography: Introduction to the special issue', *The Canadian Geographer/Le Géographe canadien*, 61 (1): 4–10. doi: 10.1111/cag.12357.

van Aalst, M.K., Cannon, T., and Burton, I. (2008) 'Community level adaptation to climate change: The potential role of participatory community risk assessment', *Global Environmental Change*, 18 (1): 165–179. doi: 10.1016/j.gloenvcha.2007.06.002.

von Benzon, N. and van Blerk, L. (2017) 'Research relationships and responsibilities: 'Doing' research with 'vulnerable' participants: introduction to the special edition', *Social & Cultural Geography*, 18 (7): 895–905. doi: 10.1080/14649365.2017.1346199

Watts, J.H. (2011) 'Ethical and practical challenges of participant observation in sensitive health research', *International Journal of Social Research Methodology*, 14 (4): 301–312. doi: 10.1080/13645579.2010.517658.

Webber, S. (2016) 'Climate change adaptation as a growing development priority: Towards critical adaptation scholarship', *Geography Compass*, 10 (10): 401–413. doi: 10.1111/gec3.12278.

Winchester, H.P.M. (1996) 'Ethical issues in interviewing as a research method in human geography', *Australian Geographer*, 27 (1): 117–131. doi: 10.1080/00049189608703161.

4 Childhood is a foreign country? Ethics in socio-spatial childhood research as a question of 'how' *and* 'what'

Kathrin Hörschelmann

Introduction

Ethical concerns have occupied children's geographers and other social researchers of childhood strongly over the last two decades (cf. Alderson 1995; Barker and Weller 2003a; Farrell 2005; Matthews 1998; Morrow and Richards 1996). This rise in attention has been closely tied to shifts in theoretical conceptualisations of childhood prompted by the new *Social Studies of Childhood* (cf. James and James 2004) and by political and legal changes codified in the 1989 UN Convention on the Rights of the Child, which all states except Somalia and the United States have ratified.

In other disciplinary contexts, such as the medical sciences, concerns about negative consequences of research on children and their exploitation have also given rise to stricter regulations, including sanctions on research that shows few demonstrable benefits for child participants. Thus, in the legal contest of *Grimes v Kennedy Krieger Research Institute, Inc* (August 2001), a court in Maryland (USA) ruled that parents "could not consent to their minor children's participation in research that posed even a minimal risk of harm if it offered no prospect of direct medical benefit to the subjects" (Mastroianni and Kahn 2002, p. 1073). The ruling shone a light particularly on the exploitation of economically and socially marginalised communities and had implications beyond the confines of medical research. As explained in the final verdict:

> [P]arents, whether improperly enticed by trinkets, food stamps, money or other items, have no more right to intentionally and unnecessarily place children in potentially hazardous nontherapeutic research surroundings than do researchers. In such cases, parental consent, no matter how informed, is insufficient.
>
> (ibid.)

While children's rights to protection from harm are at the forefront of such political and legal debates, social researchers, including children's geographers, have been particularly concerned to examine the ethical consequences

DOI: 10.4324/9780429507366-4

of participation rights that are equally enshrined in the UNCRC and that connect with new conceptual understandings of children as experts of their own life worlds (Bell 2008; Morrow 2009). The UNCRC states in Article 12 that "the child who is capable of forming his or her views" should be assured "the right to express those views freely in all matters affecting the child" and in Article 13 that children should "have the right to freedom of expression" (Unicef 1989; also see Skelton 2007). A child is defined in Article 1 of the UNCRC as "every human being below the age of 18 years unless under the law applicable to the child, majority is attained earlier." The flexibility in definition that is indicated in this formulation has become a particularly contested issue, for instance regarding different states' approaches to children's voting rights, military recruitment or involvement in armed combat. There are also stark differences between states in the definition of the age of criminal responsibility which in some states, such as the United Kingdom, is as low as 10 years (cf. Weller 2006). These issues all indicate that excluding children because they have not yet reached the formal age of maturity is fraught with contradictions that researchers need to tackle in the interest of children.

The shift in perspectives that Articles 12 and 13 in the UNCRC exemplify inevitably raises ethical questions about how best to ensure that children have their voices heard. As explained by Christensen and Prout (2002, p. 478):

> … new ethical issues inevitably arise from seeing children as social actors. The theoretical orientations through which social science constitutes children bring with them ethical implications. Once children are seen as social actors, a more complex field emerges in which there is greater scope for ethical dilemmas and new responsibilities for researchers.

Over the last two to three decades, childhood researchers across the social sciences have developed a wealth of proposals for new ethical and methodological practices that respond better to these challenges and that also consider the diversity of children's interests and needs, which can be age related but are neither necessarily nor exclusively so.

In this chapter, I first present an overview of these proposals and of the problems they address. In addition to mapping some of the practical suggestions and methodological innovations adopted by social and geographical researchers in the field of childhood studies, I draw particular attention to the tensions between children's rights to protection and participation, where participation means children's right to have a say on matters that concern them, including the terms on which they take part in research and how their data is used. As Skelton (2008) has pointed out, there is an uneasy relation between participation and protection, reflected in the UNCRC requirement to act in the best interest of the child. Who decides what is in the best interest of the child? What rights do children have to decide for themselves? On what basis (if any) should limits on children's decision-making rights be defined? These are thorny questions which point to debates that go far beyond specific concerns with the ethics of research with children. They are tied to questions of institutional and generational power as well as to conceptualisations of age, social relations and subjectivity.

Like anyone else, children live their lives in webs of social relations that connect them to others, thus making it problematic to talk about the rights of children as if they were disconnected from the rights, obligations and expectations of others (Edwards and Mauthner 2012). While researchers of childhood today mostly make the participation rights of children their starting point for judging the legitimacy of their research interests and practices, the question of protection and of respect for the responsibilities and rights of parents, other guardians and/or institutional gatekeepers and carers remains a sticking point. The responsibilities and rights of the latter can conflict significantly with the aim of respecting and prioritising children's voices and decision-making powers, as I show in the second section of this chapter.

This troublesome ethical conundrum relates closely to the topic that I will turn to in the third section of the chapter: domestic child abuse. Research ethics here concerns more than just our empirical practice. It extends to our conceptual lenses, paradigms and thematic foci. Disciplinary limits on our knowledge and expertise can easily lead to the exclusion of certain childhood experiences by default and since adults usually need to be asked for consent before children in their care can participate in research, the issue of abuse may receive limited attention due to both, researchers' conceptual blindfolds and practical ethical requirements. These problems should not, however, legitimate problematic oversights in our research. I argue that researchers in the social and spatial sciences have an ethical obligation to seek ways of compensating for missing conceptual, empirical and methodological expertise, going beyond their own disciplines where needed, so that significant aspects of children's lives are not marginalised in scholarship and, by implication, in policy making and public debate.

While we can never be fully prepared for all eventualities, there is much that can and should be done to anticipate sensitivities and emotional issues that our research may raise, in research across different age groups. I suggest some ways of doing so in the closing sections of this chapter.

Research ethics: Principles, practices and tensions

When Alderson (1995) and Morrow and Richards (1996) first wrote their agenda setting interventions on the ethics of social research with children, they were faced with a major lacunae of empirical work that included children as active participants and of any detailed, reflexive discussions on the ethics of conducting social research with children. Morrow and Richards

(1996) argued that this was largely due to children being seen as the pur-vey of psychological rather than sociological research and childhood being understood as a phase of immaturity, development and socialisation into adult roles and norms, where children were mostly researched in terms of the adults that they were becoming (or were apparently at risk of *not becoming*, cf. Jones 2009). Children were mostly seen as subjects whose experiences of the social world could not be reliably captured through their own accounts and with standard social research methods. As Morrow and Richards (1996, pp. 92–93) explained:

> [...], sociology as a discipline has tended to ignore children, and left them to psychologists to study [...] Further, sociology of the family or education often uses adults – parents or teachers – as informants about children, so even where children are the central concern of research, they may not be directly involved.

Challenging scientific paradigms that marginalise children's views and that consider their social worlds as less relevant than the norms, identities, roles and perspectives of adults was a crucial ethical step for social research, entailing not just changes in the 'how' but also in the 'what' of research. Such changes were necessary in order to create space for children's views in social and geographical research in the first place.

Reconceptualisations of children 'as already competent participants in their everyday worlds' and of childhood "as socially constructed within spe-cific times, places and contexts" (Farrell 2005, p. 6), led to new questions about ethical *practice* in research with children. Researchers increasingly questioned not just whether they, as adults, could ever 'see like a child' (cf. Jones 2001), but also how power differentials between children and adults affect social research and what can be done to counteract them (Gallacher and Gallagher 2008; Morrow 2009). Even as children have become increas-ingly recognised as competent agents in their own life worlds, social researchers of childhood emphasise that those life worlds are neither all the same, nor are they closed off from the world of adults. Childhood itself is, as a category, constructed in relation to adulthood, and when researching children's perspectives and experiences, we need to recognise that the latter is formed relationally, in everyday contexts of generational power as well as in research situations that involve power relations between adult academics who work within/for institutions and child participants who contribute their time, effort and knowledge to the benefit of others. As Morrow explains, no matter how research is approached, critical reflection on differential power relations is therefore necessary:

> The 'power' to choose which theoretical standpoint, or way of under-standing children, lies with the researcher. The research populations studied, the methods used, and crucially the interpretation of the data

collected, are all influenced by the view of children taken, and there are obvious ethical implications to this (...).

(Morrow 2009, p. 52)

Critical reflection on power differentials needs to include the dominance of adults in ethical review processes and the development of ethical guidelines. Currently, it is rare for ethical review boards to include children and the diverse ethical understandings of children are not generally taken into account when ethical protocols are developed and reviewed (cf. Skelton 2008). Here, it is still assumed that adults know best and the regulatory structures that prevail reflect generational hierarchies more than the views of children. Children can and do reflect critically on the ethics of research, however (cf. Punch 2002), and although no child can speak fully on behalf of any other child neither can adults compensate for the views of children on issues such as age-related power and resulting needs and competencies.

Including children in ethical review boards would be one step towards establishing greater 'ethical symmetry' in research that involves children and that affects their interests. As proposed by Christensen and Prout (2002, p. 482), 'ethical symmetry' between children and adults means,

> [...] that the researcher takes as his or her *starting* point the view that the ethical relationship between researcher and informant is the same whether he or she conducts research with adults or with children. This has a number of implications. The first is that the researcher employs the same ethical principles whether they are researching children or adults. Second, that each right and ethical consideration in relation to adults in the research process has its counterpart for children. Third, the symmetrical treatment of children in research means that any differences between carrying out research with children or with adults should be allowed to arise from this starting point, according to the concrete situation of children, rather than being assumed in advance.
>
> Thus, from this point of view, researchers do not have to use particular methods or, indeed, work with a different set of ethical standards when working with children. Rather it means that the practices employed in the research have to be in line with children's experiences, interests, values and everyday routines. (original italics)

What this means is that adult researchers inevitably enter ethical relationships with participating children that evolve around the same *underlying* issues as those with adults, namely the need to protect participants from harm, to respect their needs, and to act justly, with honesty and integrity (i.e. to establish reciprocal relationships, where participants derive benefits from the research). If adults are assumed to have rights regarding these key issues, then so will children. What it does *not* mean is that adult and child interests or needs are the same. Rather, the aim is to create conditions in our research that enable participating children to raise their *specific* interests, needs and concerns and for researchers to respond ethically, in *specific* contexts. This is crucial because children bring varied competencies as well as interests and needs to any research encounter. For ethical review boards, this also means a need to develop child-friendly processes and procedures as and when children are included and to reflect on ways of counteracting power-imbalances between children and adults (cf. Holt 2004). It may, for instance, be beneficial to conduct consultations with children in a separate environment and with an experienced, specialist moderator, depending on institutional cultures, participating children's preferences and the specific issues that are to be discussed. Processes and procedures should take account of intersectional differences too and consider the capabilities and social positioning of different children. Just as in research, it may be necessary to make adjustments, if the intended consultation methods do not appear to suit the actual situation, if the moderator notices negative power dynamics, or if participating children raise concerns. For, while general expertise with age-related developmental abilities can be helpful to plan for appropriate child-friendly methods, they do not account for actual diversity and can reinforce exclusions if applied unreflectively (cf. Abebe 2009).

Relational ethics: situated reflexivity, principles and practices

Here, as in other areas of research, ethics is a matter of judgement, reflexivity and negotiation in specific contexts that cannot be fully regulated in advance (Edwards and Mauthner 2012; Morrow 2009). *A priori* principles, while helpful for guiding practice, can easily conflict when they are applied to parties with diverging interests and positions, thus requiring the reconciliation of conflicting needs and demands. This is recognised by relational rather than deontological approaches to ethics, which start from the premise that responsibilities arise from relational interactions in context. As Edwards and Mauthner (2012, pp. 23, 25) explain: "What is moral and ethical is arrived at through an active and situationally contingent exchange of experiences, perspectives and ideas across differences," making ethics "about *how* to deal with conflict, disagreement, and ambivalence rather than attempting to eliminate it" (original italics). Such an approach to ethics is pertinent to research with children, where researchers may be faced with conflicting interests and needs that arise from children's situated positionalities and relations with adults, including researchers themselves. The ethical decisions that we take affect which childhood experiences and understandings we can account for in our research and, more importantly, they affect how children experience their social positioning.

Despite this theoretical emphasis on situated researcher judgments, in practice, many child researchers veer on the side of caution and assume some differences between children and adults, especially concerning vulnerabilities related to different developmental and generational statuses.

None of these apply equally to all children, in all age groups, and none of them pertain essentially only to children. However, it would be disingenuous to ignore the possibility that children may have specific requirements *qua* their developmental and generational status and it can be dangerous to ignore developmental vulnerabilities that may place children at greater risk of harm or that affect how they participate in the research. If such potential vulnerabilities are not anticipated and sought to be assessed before embarking on research with children, adult researchers risk failing in their duty of care. This is a problem that is arguably not sufficiently addressed in the literature on relational research ethics and in discussions that focus primarily on children's competencies. In order to be able to respond sensitively and appropriately to different interests and needs, researchers need to have knowledge and expertise that allows them to make reasoned, cautious decisions about potential harms to participants and, therefore, ethical requirements. This does not absolve them from the need to remain reflexive throughout the research, to address power imbalances, and to adapt to unpredictable situations and demands. However, just as with sensitive research involving adults, it means that it is often necessary and advisable to seek guidance from expert literatures in relevant disciplinary fields before we set out to conduct the research, such that predictable risks can be assessed, counteracted and avoided. Social researchers face difficulties here, as expertise from other disciplines may, (a) be difficult to access without requisite training in that discipline, and (b) follow completely different paradigms, sometimes running counter the recognition of children's competencies and diversities that is central to relational perspectives and the new Social Studies of Childhood. In practice, ethics in research with children, therefore, means a difficult balancing act between the need for prior expertise, critical reflection on that expertise, and situational responsiveness.

The issue of situatedness and ethical reflexivity also relates closely to geographical concerns. In most global contexts, children's lives are heavily institutionalised. While there are significant differences in terms of children's autonomy and their highly differentiated social responsibilities, which may include responsibilities for generating income or taking care of siblings and/ or parents (cf. Evans 2011), their everyday geographies tend to be strongly impacted by adult and institutional rules and norms. This means, plain and simple, that researchers rarely interact with children in settings that are free from the supervision and interference of other adults. Ethical practices have to conform with institutional practices in order to gain permission for research in schools and childcare settings and often need to be negotiated with parents, legal guardians and adult professionals. While the main reason for this also an ethical concern, namely the need to protect children against harm, it rubs against the participation rights of children and can conflict strongly with researchers' ethical commitment to include children on their own terms. Here, research with children differs significantly from that with adults, as the socio-spatial construction of their everyday lives is tied to age-related social hierarchies and legal provisions. The boundedness of children's geographies and increasing restrictions on their autonomy further means that some of their experiences and their context-specific perspectives are hard to access, even if this would be important in order to account for particularly problematic issues such as power dynamics, exploitation and abuse in domestic and institutional spheres. I return to this issue below but raise it briefly here to draw attention to the ethical dilemmas that face researchers not only in terms of their conduct (the *how* of research) but also their subject matters (the *what* of research).

Consent

Children's geographies also relate to the thorny issue of informed consent. The need to seek children's own consent and to do so honestly and fully has become widely recognised and enshrined in legislation. It is most certainly no longer sufficient to obtain permission from adult guardians:

> When participating in research children need to be informed in such a way so that they understand their own right to decide whether they wish to take a part and they are enabled to make independent choices and contribute to research design throughout the research process.
> At the same time, however, a research practice based in a set of ethical values requires the active consideration of children as fellow human beings and a continual sensitivity to their own emotions, interests and considerations in the varied situations of their lives.
>
> (Christensen and Prout 2002, p. 493)

Such a move is coherent with the recognition of children's agency and with a critique of generational power relations, but it has implications for research ethics that often link back to recognition of children's different competencies and understandings. Thus, information about the aims, risks and benefits of research projects needs to be presented clearly and in such a way that participating children can understand what they are consenting to or not (cf. Morrow 2009; Punch 2002). Before they can give informed consent, children who have not learnt to read or write will also require the presentation of information about the research in different formats. In order to address this issue, researchers such as Thomas and O'Kane (1998) have used audio recordings of key information about their projects in addition to written forms or information sheets.

Further, children's consent needs to be obtained without pressure or bribery. While this may seem noncontroversial and straightforward, in practice, it is easy for researchers to fall foul of this requirement (cf. Gallagher et al. 2010). Thus, they may not always be sure how far to go in outlining risks and what prior knowledge (child) participants bring with them. Does this prior knowledge suffice for them to understand the full implications of the research? Are there misunderstandings or perhaps even unwarranted concerns resulting from participants' prior knowledge that the researcher ought to address? And at what point does a reward for participants' efforts become a bribe, an attempt to persuade children to consent? The fact that the success of research projects often depends crucially on the recruitment of participants means that, in practice, researchers may be tempted to apply subtle (and not so subtle) pressure on participants. It takes vigilance and self-awareness to avoid this and we must at all points be alert to the possibility that participants may feel emotionally blackmailed into consenting. For instance, they may not wish to disappoint a researcher whom they like personally or where they feel that the researcher will be in a difficult situation if they decline participation. The generational power relations mentioned above become an additional stumbling block here. No matter how often we reiterate that participation is voluntary, children may not feel that they are allowed to say 'no'. The dependent relationships with adults in which children mostly find themselves may impact on whether they feel able to give or withhold their consent freely.

A further, extremely tricky issue is that of adult carers' rights to have their consent sought too. One of the key ways in which ethical review boards seek to minimise harm to children in research is to require the consent of parents or of adults acting *in lieu parentis*, such as teachers or child carers (cf. Barker and Weller 2003b; Morrow 2009; Skelton 2008). As adults generally have safe-guarding obligations toward children in their care, denying them the opportunity to give consent would mean placing them in difficult situations, where researchers effectively ask them to renege on their responsibilities. Neither is this legally permissible in most contexts nor do the ethical reasons for doing so strike me as sufficiently weighty to support such a practice. However, the matter is not always easily resolved. Thus, if parents or other guardians refuse to give their consent *against* the wishes of the child, it can put researchers and their potential participants in an extremely difficult ethical position. Skelton (2008, p. 27) has pointed out that ethical guidelines which "insist on adult permission and consent before a child can participate in research" contravene the UNCRC (also see Bell 2008):

> It leaves children locked within the authority of parents/guardians and unable to make a decision for themselves about their own involvement in research which specifically pertains to an aspect of their own lives experience. What if the child really wants to participate but the parent says no? Legally we would probably have to side with the parental decision but ethically we would probably want to go with the child's choice.
>
> (ibid.)

One step that researchers can take in these situations is to engage further with the carers or gatekeepers, to provide more information, answer questions, and perhaps put additional safeguards in place. Engagement and negotiation here can have the positive effect of responding better to the needs of all partners potentially affected by the research. It recognises that research often intervenes in wider relationships and provides an opportunity to minimise harmful consequences of such interference. Thus, children may wish to (or be asked to) report on their everyday experiences in ways and formats that allow insights into the lives of adults, siblings, friends and others who have not been as fully informed about the research and who become its subjects by default. In my own research, this was the case for instance with photographs that participants were asked to take of important places in their everyday lives (Hörschelmann and Schäfer 2005, 2007). A high number of participants also photographed family members and friends with whom we had never conversed and whose knowledge about the research (i.e. why they were being photographed) came only from participants themselves, if at all. In interviews, it is also common (and often wanted) that participants talk about significant others in their lives. This is a general dilemma, and not one that only pertains to research with children, but either way it means that adults (and others) may well be justified to expect that their consent is sought. Children's rights here cannot necessarily be assumed to take priority over those of others.

Virginia Morrow (2009) further emphasises that seeking consent is an ongoing process, as participants learn more about the research in the course of taking part in it and, as a result, may have new questions or change their minds. She also reminds researchers of their obligation to stay honest with participants regarding the *benefits* of research to them, about how *confidentiality* will be handled if researchers suspect a child protection issue, and about potential *uses of the data*. Concerning the benefits of the research, with the exception of participatory action research projects, it is often the case that our research findings will not lead to significant *immediate* changes and that the benefits accrue more to academic audiences and to wider publics via influences on political decision making than to participants themselves. It is crucial to make this clear in our communication with participants such as not to mislead and raise false hopes. Alternatively, many childhood researchers today advocate participatory approaches that enable children to draw more direct benefits from the research, such as tackling a particular issue, learning more about one another, acquiring skills or contributing directly to data analysis and dissemination (van Blerk and Ansell 2007; Cahill 2004, 2009).

Morrow (2009) also warns that agreements may need to be broken by researchers if they suspect a child protection issue. Participants need to be made aware of this prior to obtaining consent. Morrow advises to speak to participating children first, should this situation arise and discussing with them how best to respond. However, she also acknowledges that in some contexts, this may not be possible or appropriate. My own research has always been conducted in institutional contexts such as schools or youth clubs that provide expert support so that problematic issues could be raised with a teacher or social worker. I have not, in practice, been confronted with the need to report on any issues that the specialists were not aware of already. However, in order to be honest with participants and able to act on my duty of care, I have consistently opted to inform participants that I or the researchers working with me would speak to a nominated teacher or social worker if we were concerned about any issues that we felt might place the participant at risk of harm. In such cases, we would have talked to the children concerned first, however, to ensure that the person we would inform has their trust. A further precaution is to provide information on counselling and other support services such as telephone helplines to participating children.

Birbeck and Drummond (2007) further advise against researchers adopting passive observer roles in research, e.g. in schools or child care contexts, and to become active participants with clearly defined roles instead. This means that, when necessary, adult researchers may intervene in a research setting if they witness situations that could cause harm to some or all participants. For the authors, it is crucial that adult researchers provide an environment in which child participants can feel safe, supported and valued. Children can only participate meaningfully if this is the case. In Australia, the National Statement on Ethical Conduct in Research Involving Humans (2007, updated National Health and Medical Research Council 2018, p. 66) thus stipulates that "[t]he circumstances in which the research is conducted should provide for the child or young person's safety, emotional and psychological security, and wellbeing." As children live in socially defined relationships with adults, they will and should be able to expect that adult researchers take on duties of care normally expected of them. This means that they cannot remain completely detached, impartial observers, but may have to intervene in situations that may cause distress. Such interventions do not need to remain one-sided. They are also an opportunity for children to raise their opinions and to reflect on social power relations.

Finally, it is important to clarify how the data gathered during the research will be used. It is generally difficult to explain fully how academic data analysis and knowledge dissemination work, how widely expertise will circulate and exactly how data (such as interview excerpts) may be used in publications. Given current moves towards open data access and archiving, researchers themselves often cannot know for certain how others will potentially work with the data that they have gathered. Making such uncertainties explicit is certainly necessary but it is not sufficient, in my view, in light of the need for *informed* consent and reciprocal relationships. Even with anonymisation, open-access data archiving here stands in stark contradiction to the ethical symmetries that researchers may seek to achieve. When data that has been gathered through careful relationship building is made available to researchers that are unknown to participants, the relational aspect of data interpretation and analysis is lost and participants may feel their trust betrayed as their consent for data use arose from the process of building rapport and contextualisation that has not been established with other data users. The risk of exploitation and breeches in trust has not been sufficiently considered by funding agencies and academic publishers yet, in my view.

Diversified methodologies

A further way in which researchers have sought to respond more appropriately to the specific ethical issues raised by research with children has been a broadening of methodological approaches. Strong cases have thus been made for the benefits of child-centred and participatory methods (Barker and Weller 2003a, 2003b; Cahill 2004, 2009; Morrow 2009; Punch 2002; van Blerk et al. 2009). Creative techniques such as participatory diagramming, drawing, photography and/or video diaries have been argued by many researchers to respond particularly well to the competencies of children, drawing on skills that they often have because of age-related socialisations and/or allowing children to participate who would feel intimidated or inhibited by more traditional survey or interviewing methods (Thomas and O'Kane 1998). A consensus has emerged among researchers using these methods that they are better suited to levelling power differentials between child participants and adult researchers and that they hold the potential to be more inclusive (cf. Barker and Weller 2003b; Morrow 2009; van Blerk et al. 2009; Young and Barrett 2001). As explained by Thomas and O'Kane: "Although concrete, the activities [enable] children to talk about complex and abstract issues and to interpret the social structures and relationships that affect their lives" (1998, p. 342).

Whether this view is warranted and to what extent it risks reproducing generational hierarchies, stereotypes about children and norms of behaviour has been a matter of some debate, however (Gallacher and Gallagher 2008; Thomson 2009). Thus, Gallacher and Gallagher ask "whether 'participatory methods' can deliver all that they promise" (2008, p. 499) and warn that participatory research may in itself entail requirements for certain behaviours – fitting into prescriptive pedagogies – that run counter to their empowering and emancipatory aims.

I have personally often been troubled by the assumption that visual and creative methods are more appropriate for research with children because they are seemingly more 'fun' (cf. Barker and Weller 2003b). Not only does this underestimate how challenging and intellectually demanding it can be to, for instance, abstract aspects of social life on a map, a drawing or a photograph. It also makes vast assumptions about children, who may in fact not be familiar with some of these methods (cf. Punch 2002) or who may not feel competent using them (e.g. because of differing fine motor skills). And although adults may be less willing to use the same methods, it does not follow that they would be any less suitable for research with them.

All of the issues discussed above point to the need for researchers to remain reflexive throughout their research, to attend carefully to differing and emerging needs of participants in their specific contexts (including their relations with other adults), and to prepare well for the range of situational specifics that they may find themselves confronted with. Such preparation includes vigilance against, and efforts to counteract, the marginalisation of some children's lives and geographies in the literature, such as to gain sufficient expertise to minimise harm to children no matter what their needs may be. The following section highlights one such marginalisation and proposes some tentative ways to respond to it.

Domestic child abuse

There is a noticeable absence, compared to the breadth and depth of other research on children's socio-spatial lives, of the issue of domestic child abuse and adults' maltreatment of children in geographical scholarship. When scanning the literature on children's geographies, in particular, we find few engagements with this deeply troublesome aspect of numerous children's lives. As psychological and medical research has shown, domestic child abuse and maltreatment continues to be a significant problem across the globe, however (D'Andrea et al. 2012; Gilbert et al. 2009; Jackson et al. 2015; Katz and Barnetz 2014; van der Kolk et al. 2005; Witt et al. 2018). Thus, in Germany, it is estimated that around one in three adults have experienced violence as a child and every seventh to eighth was sexually abused as a child (Witt et al. 2018). According to D'Andrea et al. (2012), globally, around one-third of children are physically abused, while a prevalence study of child maltreatment in the United Kingdom by Radford et al. (2011) found that ca. 15.5% of 11–17 year olds and 24.1% of 18–24 year olds experienced contact and non-contact sexual abuse during childhood (cited in Jackson et al. 2015, p. 323).

Given the scale of the issue of child maltreatment and abuse in domestic contexts, and its major long-term impacts on survivors (cf. Finkelhor et al. 2005; Hörschelmann 2017; Willis et al. 2016) its relative absence from geographical childhood research and the limited amount of work on children's lived experiences of abuse (cf. Jackson et al. 2015; Willis et al. 2015) is concerning, if not necessarily surprising. Ethical concerns about the potential for harm that research on child abuse may cause are no doubt a major and very valid reason. These concerns are particularly relevant for disciplines that have little practice and expertise in conducting research on this sensitive subject. Silencing the issue completely, however, leads not only to conceptual problems with understandings of childhood diversity. It restricts possibilities for supportive interventions and for political actions that would benefit affected children and adult survivors. As Becker-Blease and Freyd (2006, p. 223) explain:

> Just as researchers underestimate the benefits of asking about abuse, they underestimate the risk of *not* asking. When we do not ask, science and humanity lose important information (Freyd et al., 2005). Further,

they withhold child protective services responses that prevent future harm. We deprive participants of the opportunity to learn about normal reactions to abuse and about community resources that could help. Studies that ask about abuse help break the taboo against speaking about abuse, helping survivors to know that talking about their experiences is important.

In order to juggle these conflicting issues and researcher responsibilities, I would make three preliminary proposals for research with children, in geography and beyond. My first proposal concerns *research practices*: ensuring adequate training for those conducting research with children (also see Becker-Blease and Freyd 2006). This may seem obvious but is, in my experience, often neglected in debates on research ethics. The training should include thorough knowledge about the prevalence, forms and impacts of child maltreatment and abuse and, ideally, practical counselling training so that we can plan our research and respond as adequately as possible to prevent harm to children who are most at risk. We should, ideally, be personally qualified to offer appropriate support as and when needed. If this is not possible, then it is preferrable in my view for the research to be conducted within an institutional environment that provides the necessary support. I would acknowledge, however, that this places limits on the kinds of research that we can conduct and that there are, thus, trade-offs that need reflection and discussion within our disciplines.

As acknowledged above, the knowledge and expertise that is required in order to be adequately prepared will often need to be interdisciplinary, especially where our own disciplinary knowledge is limited. While this means recognising and seeking to understand other knowledge traditions, it does not mean abandoning our own conceptual lenses but putting them work so that we can draw on other knowledges while reviewing them in light of critical perspectives that stem from our own disciplinary approaches. To give just one example: from a social constructionist perspective, it is crucial to question how the frequent labelling of persons and practices as 'deviant,' 'oppositional defiant' or 'dysfunctional' in medical psychology contributes to social stigmatisation, discrimination and the (re)production of problematic power relations.

The second proposal that I would make concerns the *knowledge* that we include in our *analyses and representations* of children's social worlds beyond that which we have gained from our own empirical research. Here again, interdisciplinary research can help us to fill in some of the blanks, or at least to show more clearly where the limits of our own knowledge may lie. It is neither realistic nor desirable for all childhood researchers to address the issue of maltreatment and abuse. Nor does this necessarily make sense. There are clearly areas of research where this issue is likely to be less of a concern. And, research that explicitly addresses it requires *specialist* skills in order to minimise harm. However, we can and should make rigorous use of research from other disciplines where this is necessary to ensure that we do not reproduce problematic blind-folds in our analyses and write-ups. For me, the problem is less that little geographical research has been conducted in this area and more that we too often neglect the knowledge that does exist, thus producing a lop-sided understanding of the diversity of children's social worlds.

My third, and wider, proposal is to heed the learnings of research on the prevalence and long-term consequences of child maltreatment and abuse *more generally*, so that we may, (a) broaden our conceptualisations of social relations, personhood and social geographies, and (b) ensure ethical research practices that are more responsive to the diversity of personal geographies which we encounter in our research. These personal geographies include those of *adult survivors*. It is our responsibility to respond sensitively to their needs in empirical research *as well as* to make visible their accounts and experiences as an aspect of differentiated social geographies (cf. Willis et al. 2015). Given that some researchers and students have themselves been affected, we also have a duty of care that obliges us to be more fully aware and better informed. It is for this reason too that I consider research ethics as a question of '*what*' as well as '*how*'.

Conclusion

In this chapter, I have reviewed some of the ethical and methodological practices that have been developed in social and geographical research with children in order to respond better to changes in conceptualisations of childhood and in children's rights, engaging particularly with the tension between participation and protection and with the question of *what* to research in addition to *how*.

The chapter has underlined Christensen and Prout's (2002) contention that childhood researchers need some ethical principles and guidance, but that these have to be applied in concrete situations and adapted. As the authors have argued, we also need more dialogue among scholars and with children on research ethics:

> While both general codes of ethics and individual responsibility are required, it is essential that these be linked through a wider discussion of ethics that is placed more centrally in the professional activities of childhood researchers. A dialogue with children throughout the research process, we suggest, is an essential component of this work.
>
> (ibid., p. 478)

One particularly crucial issue to continue reflecting on in our scholarly debates and research practice is that of generational power relations. It could be argued that this issue indeed underlies most of the ethical concerns that relate specifically to research with children.

However, *generational* power creates a specific additional dimension that needs to be considered, both as a wider issue that researchers may seek to challenge through their research and as an intricate aspect of research relations themselves. It touches also on the questions of vulnerability and protection that were addressed in the last section of this chapter. Further, just as adults cannot simply turn back the clock to their own childhoods or assume that their memories of childhood are relatable to those of the children they are researching with, so they cannot assume to know how different participants experience relations of power in their research projects. Engaging with children's views and creating research conditions in which participants feel able to articulate concerns (which researchers will take seriously and respond to) is crucial for research to be conducted more ethically in this respect. A further balancing act for researchers is negotiations with adult carers and gatekeepers. Here, the need to consult with adults in the interest of preventing harm to child participants must be weighed carefully against principles of confidentiality, participation rights and children's wishes to have their voices heard (Skelton 2008).

The ethical dilemmas arising from this relate particularly to the issue addressed in the final sections of the chapter, which drew attention to *child abuse and maltreatment* as matters not just to the *how* but also the *what* of research. I have tentatively proposed some ways forward, making suggestions for ethical practice and outline possibilities for closing the gap in understanding by getting adequate training and drawing on research in other disciplines. Research with adult survivors is another possibility, if not for researching children's experiences then for understanding longer-term impacts on personal geographies better (cf. Willis et al. 2016).

I have argued in this chapter that the exclusion of domestic child abuse from scholarship on, and beyond, children's geographies and social worlds needs to be challenged not least because we otherwise risk conducting empirical research without sufficient preparation. Lack of interdisciplinary knowledge and expertise can lead to significant harm to research participants (including adult survivors of child abuse and researchers themselves). Interdisciplinary knowledge can compensate to some extent for gaps in our disciplinary understandings that we do not feel equipped to fill otherwise. Research with traumatised children and abuse survivors requires skills that few human geographers have acquired as part of their training and there are therefore strong ethical reasons for abstaining from it. Here, I have made the case for reflexively reengaging with knowledge from other disciplines, including that which human geographers have largely distanced themselves from due to concerns about problematic social constructions, essentialisms and universalisations, in developmental psychology for instance. I make the case for a cautious reengagement *because*, rather than despite, of ethical concerns. Developmental psychology, public health and medical research deliver insights into child wellbeing that support the aim of tackling the marginalisation and exclusion of certain childhoods, while the skills of deconstruction and knowledge of children's agencies that social researchers and children's geographers bring to the table can, and should, inform such literatures in turn.

References

Abebe, T. (2009) 'Multiple methods, complex dilemmas: Negotiating socio-ethical spaces in participatory research with disadvantaged children', *Children's Geographies, 7* (4): 451–465.

Alderson, P. (1995) *Researching with children: Children, ethics and social research.* London: Save the Children.

Barker, J. and Weller, S. (2003a) 'Geography of methodological issues in research with children', *Qualitative Research,* 3 (2): 207–227.

Barker, J. and Weller, S (2003b). '"Is it fun?" developing children centred research methods', *International Journal of Sociology and Social Policy,* 23 (1): 33–58.

Becker-Blease, K. A. and Freyd, J. J. (2006) 'Research participants telling the truth about their lives: The ethics of asking and not asking about abuse', *American Psychologist,* 61 (3): 218–226.

Bell, N. (2008) 'Ethics in child research: Rights, reason and responsibilities', *Children's Geographies,* 6 (1): 7–20.

Birbeck, D. J. and Drummond, M. J. (2007) 'Research with young children: Contemplating methods and ethics', *The Journal of Educational Enquiry,* 7 (2): 21–31.

Cahill, C. (2004) 'Defying gravity? Raising consciousness through collective research', *Children's Geographies,* 2 (2): 273–286.

Cahill, C. (2009) 'Doing research *with* young people: Participatory research and the rituals of collective work', in L. van Blerk and M. Kesby (eds.) *Doing children's geographies: Methodological issues in research with young people.* London and New York: Routledge, pp. 98–113.

Christensen, P. and Prout, A. (2002) 'Working with ethical symmetry in social research with children', *Childhood,* 9 (4): 477–497.

D'Andrea, W., Ford, J., Stolbach, B., Spinazzola, J. and van der Kolk, B.A. (2012) 'Understanding interpersonal trauma in children: Why we need a developmentally appropriate trauma diagnosis', *American Journal of Orthopsychiatry,* 82 (2): 187–200.

Edwards, R. and Mauthner, M. (2012) 'Ethics and feminist research: theory and practice', in T. Miller, M. Birch, M. Mauthner and J. Jessop (eds) *Ethics in qualitative research.* London: Sage, pp. 14–28.

Evans, R. (2011) '"We are managing our own lives...": Life transitions and care in sibling-headed households affected by AIDS in Tanzania and Uganda', *Area* 43 (4): 384–396.

Farrell, A. (2005) 'Ethics and research with children', in A. Farrell (ed) *Ethics and research with children.* McGraw-Hill: Open University Press, pp. 1–14.

Finkelhor, D., Ormrod, R., Turner, H. and Hamby, S. L. (2005) 'The victimization of children and youth: A comprehensive, national survey', *Child maltreatment: Journal of the American Professional Society on the Abuse of Children,* 10: 5–15.

Gallacher, L. A. and Gallagher, M. (2008) 'Methodological immaturity in childhood research? Thinking through participatory methods', *Childhood,* 15 (4), 499–516.

Gallagher, M., Haywood, S. L., Jones, M. W. and Milne, S. (2010) 'Negotiating informed consent with children in school-based research: A critical review', *Children and Society*, 24 (6), 471–482.

Gilbert, R., Widom, C.S., Browne, K., Fergusson, D., Webb, E. and Janson, S. (2009) 'Burden and consequences of child maltreatment in high-income countries', *The Lancet*, 373 (9657): 68–81.

Holt, L. (2004) 'The "voices" of children: De-centring empowering research relations', *Children's Geographies*, 2 (1): 13–27.

Hörschelmann, K. (2017) 'Violent geographies of childhood and home: The child in the closet', in C. Harker, K. Hörschelmann and T. Skelton (eds) *Conflict, violence and peace: Geographies of children and young people*. Singapore: Springer, pp. 233–251.

Hörschelmann, K. and Schäfer, N. (2005) 'Performing the global through the local—globalisation and individualisation in the spatial practices of young East Germans', *Children's Geographies*, 3 (2): 219–242.

Hörschelmann, K. and Schäfer, N. (2007) '"Berlin is not a foreign country, stupid!"—Growing up "global" in Eastern Germany', *Environment and Planning A*, 39 (8), 1855–1872.

Jackson, S., Newall, E. and Backett-Milburn, K (2015) 'Children's narratives of sexual abuse', *Child and Family Social Work*, 20: 322–332.

James, A. and James, A. L. (2004) *Constructing childhood: Theory, policy and social practice*. Basingstoke and New York: Palgrave Macmillan.

Jones, G. (2009) *Youth*. Cambridge, UK, and Malden: Polity Press.

Jones, O. (2001) '"Before the dark of reason": Some ethical and epistemological considerations on the otherness of children', *Ethics, Place & Environment*, 4 (2), 173–178.

Katz, C. and Barnetz, Z. (2014) 'The behavior patterns of abused children as described in their testimonies', *Child Abuse & Neglect*, 38 (6): 1033–1040.

Mastroianni, A. C. and Kahn, J. P. (2002) 'Risk and responsibility: Ethics, Grimes v Kennedy Krieger, and public health research involving children', *American Journal of Public Health*, 92 (7), 1073–1076.

Matthews, H. (1998) 'The geography of children: Some ethical and methodological considerations for project and dissertation work', *Journal of Geography in Higher Education*, 22 (3): 311–324.

Morrow, V. (2009) *The ethics of social research with children and families in young lives: Practical experiences*, Oxford: Young Lives, Working Paper No. 53.

Morrow, V. and Richards, M. (1996) 'The ethics of social research with children: An overview 1', *Children and Society*, 10 (2), 90–105.

National Health and Medical Research Council (2018) *National Statement on Ethical Conduct in Human Research 2007 (updated 2018)*. Canberra: Commonwealth of Australia.

Punch, S. (2002) 'Research with children: The same or different from research with adults?', *Childhood*, 9 (3): 321–341.

Radford, L., Corral, S., Bradley, C., Fisher, H., Basset, C., Howatt, N. et al. (2011) *Child abuse and neglect in the UK today*. London: NSPCC.

Skelton, T. (2007) 'Children, young people, UNICEF and participation', *Children's Geographies*, 5 (1–2): 165–181.

Skelton, T. (2008) 'Research with children and young people: Exploring the tensions between ethics, competence and participation', *Children's Geographies*, 6 (1): 21–36.

Thomas, N. and O'Kane, C. (1998) 'The ethics of participatory research with children', *Children and Society*, 12 (5): 336–348.

Thomson, F. (2009) 'Are methodologies *for* children keeping *them* in their place?', in L. van Blerk and M. Kesby (eds) *Doing children's geographies: Methodological issues in research with young people.* London and New York: Routledge, pp. 183–194.

Unicef (1989) *The United Nations Convention on the Rights of the Child*, https://www.unicef.org.uk/what-we-do/un-convention-child-rights/ (accessed: 15 May 2020).

van Blerk, L. and Ansell, N. (2007) 'Participatory feedback and dissemination with and for children: reflections from research with young migrants in southern Africa', *Children's Geographies*, 5 (3): 313–324.

van Blerk, L., Barker, J., Ansell, N., Smith, F. and Kesby, M. (2009) 'Researching children's geographies', in L. van Blerk and M. Kesby (eds) *Doing children's geographies: Methodological issues in research with young people.* London and New York: Routledge, pp. 7–14.

van der Kolk, B.A., Roth, S., Pelcovitz, D., Sunday, S. and Spinazzola, J. (2005) 'Disorders of extreme stress: The empirical foundation of a complex adaptation to trauma', *Journal of Traumatic Stress*, 18 (5): 389–399.

Weller, S. (2006) 'Situating (young) teenagers in geographies of children and youth', *Children's Geographies*, 4 (1): 97–108.

Willis, A., Canavan, S. and Prior, S. (2015) 'Searching for safe space: The absent presence of childhood sexual abuse in human geography', *Gender, Place and Culture*, 22 (10): 1481–1492.

Willis, A., Prior, S. and Canavan, S. (2016) 'Spaces of dissociation: The impact of childhood sexual abuse on the personal geographies of adult survivors', *Area* 48 (2): 206–2012.

Witt, A., Glaesmer, H., Jud, A., Plener, P. L., Brähler, E., Brown, R. C. and Fegert, J. M. (2018) 'Trends in child maltreatment in Germany: Comparison of two representative population-based studies', *Child and Adolescent Psychiatry and Mental Health*, 12 (1): 24.

Young, L. and Barrett, H. (2001) 'Adapting visual methods: Action research with Kampala street children', *Area*, 33 (2): 141–152.

5 Ethical challenges arising from the vulnerability of refugees and asylum seekers within the research process

Dorit Happ

Introduction

Already prior to the often called 'refugee crisis' in Europe in 2015, uncontrolled migration was predominately perceived as a security threat to the internal stability and cohesion of addressed destination countries, within political and public discourse (see Peoples and Vaughan-Williams 2010: 136; Huysmans 2006: 46; Ehrkamp and Leitner 2006: 1592; Bigo 2002: 64). As the political perception of migration as a threat is mainly initiated by state organs and their implemented migration policy, migration studies and political geographers for a long time conducted state-centred analysis. They have focused less on refugees' and asylum seekers' perspectives and their experiences while integrating in the host country. As a consequence, critical migration studies and feminist geographers have called for shifting the perspective from the state to the individual level, i.e. to migrants themselves (see Keely and Kraly 2018: 34; Freedman et al. 2017: 11; Hyndman 2012: 244; Peoples and Vaughan-Williams 2010: 140; Ehrkamp and Leitner 2006: 1595; Gilmartin 2008: 1839).

In order to capture migrants' perspectives, ethnographic qualitative methods are often applied as they allow personal experiences and perceptions of migrants to be examined (see Beuving and De Vries 2015: 19; Chatty 2014: 1). In this vein, within the emergent field of refugee studies, small case studies including in-depth interviews, focus groups and intense fieldwork are often conducted that facilitate the assessment of patterns of practice and issues of migration *in situ*. Furthermore, on the one hand, scholars within refugee studies value qualitative methods highly in order to assess participants' needs and concerns (see Düvell et al. 2010: 231; Mackenzie et al. 2007: 307; Mountz et al. 2003: 41). On the other hand, data gathered with the help of qualitative methods is extremely personal, sometimes intimate, and needs to be kept confidential as its misuse may endanger respondents (see Düvell et al. 2010: 228). This is of special significance when conducting research with vulnerable groups such as refugees and asylum seekers. Qualitative research with refugees and asylum seekers may create "spaces in which the researched are able to speak for themselves" (McEwan 2009:

DOI: 10.4324/9780429507366-5

333). Additionally, excluding vulnerable groups from research may reinforce their vulnerability, as their voices and perspective are further excluded from scientific and public discourse (see Birman 2006: 160).

This chapter will discuss ethical dilemmas that arose when conducting qualitative research with vulnerable participants – in particular refugees and asylum seekers in Belarus and Ukraine. First, the chapter will outline the term vulnerability and its implication for research with refugees and asylum seekers. The chapter follows a constructivist approach highlighting the socially constructed and relational nature of vulnerability in contrast to a positive understanding of a 'natural' and fixed vulnerability (see Brown et al. 2017: 499). Vulnerability is conceptualised here as dependant on the context and circumstances which affect groups differently, as they are more or less vulnerable to exploitation (see von Benzon and van Blerk 2017: 3). Furthermore, the chapter discusses the risk that refugees and asylum seekers may experience vulnerability within the research process itself, which may reinforce their positioning and perception as vulnerable. The chapter will reflect on ethical and methodological implications of carrying out research with refugees and asylum seekers in light of these concerns. It is based on experiences from seven months of qualitative fieldwork in Belarus and Ukraine (2012 and 2014), when I conducted research on the integration of refugees and asylum seekers into their new host societies. These field studies are of a larger project on the EU's extraterritorial engagement in Eastern Europe, funded by the Volkswagen Foundation and led by Bettina Bruns.

Researching *vulnerable* groups – concepts and implications

The ascription 'vulnerable' encompasses different factors and characteristics related to the concepts of harm, need, power, dependency, capacity and exploitation (see Mackenzie et al. 2014: 1). Clark emphasises three key issues: physical weakness, emotional instability and economic dependency (2007: 288–291). Additionally, Emmel and Hughes (2010) conceptualise a 'social space of vulnerability' consisting of limited resources for everyday needs, lack of resilience and limited capacities to address these needs, which culminate in dependence on external help and service provision (see Emmel and Hughes 2010: 177, also von Unger 2018: 4).

The United Nations High Commissioner for Refugees (UNHCR) and the International Detention Coalition (IDC) describe reasons why refugees and asylum seekers find themselves in a vulnerable position: they "have typically experienced life-changing upheaval, danger, loss and fear, and face enormous adjustments to life in a new environment" (UNHCR and IDC 2016: 18). According to the UNHCR's definition of refugees as persons who had to migrate due to a well-profound fear of persecution in their home country, refugees are *per se* vulnerable. Due to their migration, they have lost the protection of their home country. Consequently, refugees occupy an in-between position as a "refugee has lost the protection of his country of origin and cannot yet enjoy the protection of his new country of residence" (Hovy 2018: 43). The United Nations New York Declaration for Refugees and Migrants (2016) additionally emphasises that refugees are exposed to xenophobic behaviour, racism and discrimination in the societies of their new host countries (United Nations 2016: 8). Moreover, due to their displacement, refugees experience 'multiple losses' (Ellis et al. 2007: 461), among them often the absence of their family members, supporting networks, former working place and lifestyle, and in general the protection of their civil rights in their homeland. Arriving in a transit or host country, refugees and asylum seekers still have to overcome difficult challenges connected to the process of integration, e.g. learning the local language, finding a place of residence and a working place securing their financial viability. Furthermore, the majority of refugees and asylum seekers still struggle to deal with traumatic events experienced during the migration process (see Gibney 2018: 7; Ellis et al. 2007: 465). Therefore, to sum up, vulnerability of refugees and asylum seekers is a multidimensional and dynamic process that continues after their arrival in the host countries and is closely linked to their psychological and physical health conditions.

Research on migrants' vulnerability is valuable, as it makes the challenging living conditions of refugees and asylum seekers explicit. However, labelling refugees and asylum seekers as 'vulnerable' also needs critical discussion, as it risks creating and contributing to discrimination, stereotyping and the homogenisation of this research group as passive victims in need of 'paternalistic' humanitarian help without their own agency (see Mackenzie et al. 2014: 16). Thus, NGOs and other international and national donor organisations may utilise the ascription 'vulnerable' in order to justify their engagement and to gain financial support (see Smith and Waite 2019: 2296; Marlowe 2018: 139; Brown et al. 2017: 500; Sigona 2014: 3). Furthermore, Butler et al. state that vulnerability "always operates within a tactical field" (2016: 5), where vulnerability is first ascribed to 'others' – a group or community – in order to establish and maintain one's own power. Secondly, as a paternalistic strategy, the process of claiming and attributing vulnerability is used in order to separate groups from each other. The ascription as vulnerable therefore may establish hierarchies within or between social groups. To sum up, applying a deficit perspective leads to a misguided perception of refugees and asylum seekers as passive victims. Consequently, applying the label vulnerability may also result in paternalistic and discriminatory behaviour (see Peter and Friedland 2017: 111; Baranik et al. 2018: 117). In this vein, Sabsay criticises humanitarian organisations as they initiate the "construction of 'the suffering other' as a mute and helplessly un-nurtured, violated, or deprived" group (2016: 280). Additionally, research itself and even research ethics boards may contribute to the construction or the enhancement of research participants' vulnerability (see Maillet et al. 2016: 2; Aldridge 2014: 113; Perry 2011: 901; Pittaway et al. 2010: 231; Ellis et al. 2007: 470; Mackenzie et al. 2007: 300).

In this context, McLaughlin and Alfaro-Velcamp criticise researchers who lack "proper training, supervision, and ethics oversight" (McLaughlin and Alfaro-Velcamp 2015: 28), yet conduct interviews with vulnerable participants, thus exploiting power relations and risking new traumas. They call for "an increasing awareness among researchers of participant reactions and resistance to the experience of research as exploitation" (McLaughlin and Alfaro-Velcamp 2015: 28). Although ethical guidelines attract growing attention within refugee research today (see Lewis 2016: 100), studies and publications involving refugees and migrants often still lack a discussion about concrete research design and ethical implications of the research process (see Makhoul et al. 2018: 35; Salter 2013: 56; Jacobsen and Landau 2003: 2). Feminist geographers advocate for a reflexive turn within geography, discussing unequal power relations between researcher and researched and proposing methods and practices that can reduce them, such as seeking informed consent, integrating gatekeepers, working with local field assistants, etc. (see McEwan 2009: 333; Kobayashi 2005: 36).

Refugees and asylum seekers are also vulnerable in the research process for other reasons. Thus, McLaughlin and Alfaro-Velcamp (2015) identify three basic risks that participants may experience and fear within refugee studies: First, refugees and asylum seekers often have limited access to information and knowledge of the legal system of their host country. Consequently, they may "lack tools of advocacy for themselves" (McLaughlin and Alfaro-Velcamp 2015: 34), in order to decline participation. Second, refugees and asylum seekers may agree to participate if they perceive the researcher as an authoritative person and connect their participation with the hope of receiving assistance. Third, respondents may understand the interviews as a replication of their procedure of Refugee Status Determination (RSD), thus fearing that they could potentially jeopardise their legal status (see McLaughlin and Alfaro-Velcamp 2015: 34). Because of their vulnerability, refugees and asylum seekers may distrust the research projects' aim and the researchers' independence (see Kabranian-Melkonian 2015: 718). In their assessment of the concerns that refugees and asylum seekers had about participating in research, Pittaway et al. (2010) identified the following further issues: "fear of backlash from government authorities," "false expectations of assistance from researchers," "potential for retraumatisation," "mistrust of white researchers" and "exploitation by previous researchers and journalists" (Pittaway et al. 2010: 236; also see Krause 2017:10; Birman 2006: 166). To sum up, doing research with vulnerable groups or studying vulnerability requires special responsibility on the part of the researcher. This not only includes the ongoing and intense attention given to the personal safety and well-being of research participants but also reflection on the power relations and interdependencies between participants, gatekeepers and researchers. Researchers should therefore follow the principle of 'Do No Harm' closely. This includes being attentive to the possibility that the research and the research process itself may contribute to participants' vulnerability.

Practising research with refugees and asylum seekers in Belarus and Ukraine

This chapter is based on research about the integration of refugees and asylum seekers conducted in Belarus and Ukraine over a total of seven months in 2012 and 2014. As part of the project, 29 in-depth qualitative interviews were carried out with refugees and asylum seekers and a further 28 interviews with local experts working in local NGOs or at UNHCR. Two focus groups, averaging two hours each, were also conducted with asylum seekers in 2012 and 2014. Furthermore, ethnographic participant observation was carried out in detention centres, in language courses, during cultural events at a family centre, and at the workplaces of refugees and asylum seeker (in factories, own enterprises or at various market stalls owned by refugees and asylum seekers in Belarus and Ukraine).

The vulnerability of refugees and asylum seekers in Ukraine and Belarus is of unique and multi-layered character. In addition to the vulnerability that refugees and asylum seekers experience in general (and as described in the section above), most migrants understand both countries as transit locations rather than as destination countries (see Markowski et al. 2014: 66; Migration Policy Centre 2013: 1; Shevel 2011: 36). Therefore, from the migrants' perspective uncertainty and a state of limbo characterises their time in Ukraine and Belarus as the two countries have a relatively low recognition rate (see UNHCR 2017: 3) and as the procedure to determine whether or not they may be granted Refugee Status is time-consuming, with an unknown outcome, asylum seekers find themselves in a particular 'legal limbo' (Hovy 2018: 41). They are unable to exercise the same political and social rights as local citizens. Moreover, the process of Refugee Status Determination (RSD) is also susceptible to corruption (see Kaźmierkiewicz 2011: 40). This intensifies the financial difficulties of asylum seekers. Furthermore, both countries lack financial resources to support the integration of refugees and asylum seekers into their societies, e.g. through language courses, health care, housing, etc. Although integration measures are partly conducted by international organisations and donors (see Brunarska and Weinar 2013: 11), the absence of help provided by the host government contributes to the vulnerability of refugees and asylum seekers in Belarus and Ukraine.

This brings me to the two specific issues observed in my fieldwork that I wish to discuss in more detail next. My reflections were sparked by a photograph, taken during my field studies in Belarus in 2014. At that time, while studying the integration of refugees and asylum seekers into Belarusian society, I conducted a group discussion with 11 Afghan women at a public school during the time of their Russian language courses. The group discussion was organised by an Afghan woman, whom I had already met during my first field trip in 2012. This woman, I will call her by the alias of *Fahima* in my chapter in order to guarantee her anonymity, was also the person who initiated taking a picture with all participants after the discussion. I was a bit surprised by her suggestion but as she was very enthusiastic and promptly gave a camera to the teacher in order to take the picture, I did not reject and no one else loudly disagreed. Back home, I had a closer look at the picture of the group and recognised that one woman was explicitly hiding her face under her headscarf and was looking away from the camera while the other participants were looking into the camera or into the room. It was obvious that the woman hiding her face (and possibly others as well) did not agree to be photographed. This episode raised several questions, first about the role of the researcher and the relationship with Fahima as a powerful, well-established gatekeeper, and second about the possibility of ensuring informed consent during the entire research process. Both questions will be discussed in the following sections.

The role of gatekeepers and vulnerability

Refugees and asylum seekers appear as a rather closed and 'hard-to-reach' group (Lewis 2016: 101; Birman 2006: 161). Therefore, identifying gatekeepers, trusted by community members and researchers, is of high importance in order to establish contact with potential participants and in order to build up trust between researcher and researched (see van Liempt and Bilger 2018: 276; Kabranian-Melkonian 2015: 717). Establishing contact via gatekeepers at the same time risks interference with the research sampling, however, as they may control access to participants. Pittaway et al. (2010: 233) nonetheless state that researchers are often only able to contact a refugee community with such 'patronage access' provided by internal organisation, local NGOs, community leaders. In my case, the group discussion could only be organised with the help of Fahima. Fahima had the necessary knowledge of who, how and where to invite possible participants. During the first field trip in 2012, I did not succeed in organising such a discussion, as Afghan refugee women appeared hard to locate as a group. They often avoided public appearance and acted in private spaces while taking over duties of care for children and family. Approaching Afghan refugee women for participation in ethnographic research was complicated, however, not only because of difficulties with locating and identifying participants. Afghan refugee women may also be hesitant to agree to an interview, if they are not used to talking to people outside of their own community or family.

In terms of the group discussion, Fahima appeared not only as a gatekeeper but also as a cultural insider. This is discussed by Jacobsen and Landau (2003), who refer to the ambivalent role that local helpers may play in the research process. On the one hand, they may help establishing contact and in building up trust. On the other hand, they may raise fears among participants that their information will be shared by other people in the refugee community (see Jacobsen and Landau 2003: 10; also Birman 2006: 171). Engaging and consulting local NGOs, activists and government organisation may likewise help researchers to gain access to the migrant community while they are still establishing trust. However, it also carries the risk of endangering research confidentiality as respondents may fear that researchers will report back to authorities and that their anonymity will be compromised (see Clark-Kazak 2017: 13; Kabranian-Melkonian 2015: 718). Funding bodies increasingly recognise the ambiguous role of cultural insiders and urge researchers to be aware of it. Thus, within its "Guidance note for research on refugees, asylum seekers and migrants," the European Commission calls for integrating researchers with migrant or refugee backgrounds in order to bridge asymmetric power relations. At the same time, however, the Commission warns that cultural insiders may act as gatekeepers themselves (see European Commission n.d.: 2).

As Fahima was both a gatekeeper to the refugee community and a cultural insider, the described episode showed that working with Fahima did not lead to the abolishment of asymmetric power relations but could instead be described as a shift in unequal power relations from researcher and researched to gatekeeper/cultural insider and researched. This relationship may, in fact, have been of even greater asymmetric character as, due to their vulnerability, the researched women may have been highly dependent on the gatekeeper's advice, support and help in the future. Taking the picture deepened asymmetric relations thwarting all efforts to keep research in line with ethical standards and demands.

Researchers should reflect critically on the role of gatekeepers, especially when conducting research with refugee and asylum-seeking women, as they may reinforce their vulnerability. On this issue, migration scholars stress that this particularly vulnerable research group experiences greater exposure to certain threats than their male community fellows as they are often exposed to patriarchal structures (see Habib 2018: 28; Vacchelli 2018: 187; Binder and Tošić 2003: 457). Concerning the ambivalent position of gatekeepers, researchers need to balance possible benefits and risks for participants. Working with gatekeepers and local insiders may lead to an interdependence between the latter and the researcher as well as to a feeling of mutual obligation (see Bruns et al. 2018: 126). Their gratitude to gatekeepers notwithstanding, researchers have to prioritise the perspectives and needs of research participants, especially if they belong to a vulnerable group.

This also means that researchers need to emphasise respondents' right to refuse or withdraw consent at any stage of the research process and ensure that this refusal will not lead to negative consequences from the gatekeepers' side (see Clark-Kazak 2017: 12). Therefore, researchers should minimise the gatekeepers' appearance and interference in the interview process. I would further advocate for stressing the independent and protected relationship between researchers and the researched. Even with these precautions, gatekeepers may at different stages influence the research process, intentionally or not. Offering to take a picture after the group discussion was probably a polite gesture of hospitality. Nevertheless, Fahima severely influenced the research process. The following section discusses how taking a photograph can challenge and dismantle the informed consent already given by participants.

The role of informed consent and vulnerability

Establishing informed consent is a major strategy to protect participants from harm and exploitation (see Clark-Kazak 2017: 13; Mackenzie et al. 2007: 301). It is essential in order to guarantee voluntary participation and to establish a trustful and equal relationship between researcher and participants. Three main elements constitute informed consent: first, potential participants need to obtain information about the research; second, they need to understand this information; and, third, they need to decide freely whether they want to participate or not (see Krause 2017: 10). Within scientific discourse, there is a discussion whether consent needs to be gained in writing or whether it can be given in an oral way. Many scholars state that written consent forms are beneficial, constituting a protection for the researcher and the associated institution. However, it is less so for the addressed participants (see De Wildt 2016: 60; Ellis et al. 2007: 469).

Additionally, within ethical guidelines, researchers who work with vulnerable participants need to judge "the probability and degree of physical, psychological, or social harm, as well as a greater susceptibility to deception or having confidentiality breached" (CIOMS and WHO 2016: 57). While obtaining consent, and throughout the entire research process, researchers need to consider that vulnerable groups are characterised by being "relatively (or absolutely) incapable of protecting their own interests" (CIOMS and WHO 2016: 57). Ellis et al. (2007) point out that the inability of refugees and asylum seekers to make a truly voluntary choice whether they want to participate in the research or not originates in their financial dependence, limited access to education and experience of social and political discrimination (see Ellis et al. 2007: 466; also Happ et al. 2018: 30).

Taking a photograph also needs to be considered as an ethical breech from another perspective, namely the role of the body as a "site of perceived experiences and subjectivity" (Vacchelli 2018: 186). If we consider the body as a representation of contextual, gendered, social and cultural characteristics, then the Afghan refugee and asylum-seeking women in the Belarus group had agreed to participate in the discussion after, but the issue of the picture showed that taking it was not included in this approval. The women had agreed to be interviewed but not to be portrayed with their physical appearance. Their body is most likely closely connected to their religious conviction as a majority of the Afghan women wore headscarves as well. From this, we can see that protecting participants' body from the public and respecting the need to keep it private can be part of the researcher's responsibility under the principle of 'Do No Harm.' The photograph had clearly contradicted the guarantee of respondents' physical anonymity. Further, it is not clear whether the shooting of the photo and the concrete reaction of the face-hiding woman had further implications. Thus, I do not know whether the other women recognised the woman's hiding behaviour and if so, how they would have understood it. They may have rejected the reaction to hide or on the contrary, they may have sympathised with it or even regretted that they did not hide and instead looked into the camera. This may have led to conflict within the refugee group as the women experienced superior, dominant or weaker positions between each other.

The described episode clearly demonstrates that informed consent is not a single event at the beginning of the interview but rather an ongoing process. I would therefore advocate for practising reflexivity to address ethical dilemmas in fieldwork. This could be done, for instance, by signing the informed consent form at the end of an interview, as Lewis (2016: 108) suggests. Researcher and interviewee thus have an opportunity to debriefing about the interview experience after the event (ibid.). Re-establishing informed consent (in writing or orally) as a concluding part of interviews with vulnerable groups allows the researcher and interviewee to reflect jointly on the interview process, interview setting, the role of gatekeepers and the researcher and unexpected events, which may have occurred during the interview. This includes asking former participants how they experienced the interview, the setting, the questions etc.: How did/do they estimate the value of the research project? How could the research and interview process be improved? These questions could lead to an open and critical discussion about the experienced research conditions and aims. The respondent's perceptions should furthermore inform the ongoing research process and be included in the methodological reflection to contextualise the empirical analysis. Including respondents' reflections in this manner can contribute to the empowerment of research participants and therefore may reduce their vulnerability within the research process. In order to gain informed consent in such a participatory way, Community-Based Participatory Research (CBPR) as a methodological approach can be a helpful inspiration and guiding framework. CBPR understands research as co-production, conducted in partnership between researchers and researched. Furthermore, this approach not only enhances trust between research partners but also reflects constantly on how power and respect define the relationship between them (Donnelly et al. 2019: 834).

Conclusions

The protection of research participants' well-being has to be the researchers' first priority, without a doubt. This is especially the case when respondents belong to a vulnerable group. Methodological and ethical implications of the research have to be reflected with particular care before, during and after the research process. In this chapter, I have advocated for including a transparent reflexivity within the publication of research results and, more importantly, for a communicative reflexivity with respondents during and after the research process. Discussing the concrete research conditions with participants can form the basis for respondents to express their fears and to discuss their intentions and expectations regarding the outcomes of research. I therefore advocate for ongoing reflexive practices, whereby informed consent appears as a process of continuous (re-)negotiation rather than as a one-off event that is completed at the beginning of the research (see Mackenzie et al. 2007: 307). Informed consent should be understood as a continuous practice to address the conditions of the respective method and to reflect on unexpected turns and impressions during the research process. The example of the photograph also showed that the researcher's responsibility to protect participants through informed consent may need to include not only participants' answers but also how their bodies show up in the research. As a presentation of participants' subjectivity, it plays a role in the collection of qualitative data. Ethical reflection further needs to include discussing the role of gatekeepers and their relationship with research participants. This chapter invites researchers to carefully reflect on the implications of establishing contact and trust with research participants with the help of gatekeepers and cultural insiders. Working with gatekeepers leads to a shift in power relations rather than to their deconstruction.

Methods that include participants in an ongoing process of reflection during the research process can be applied in order to establish more equal relations between researcher and researched and to negotiate the roles that gatekeepers may play. Thus, CBPR aimed at encouraging respondents to participate actively and to deconstruct unequal power relations at the same time appear particularly suitable for this (cf. von Unger 2018: 13; Aldridge 2014: 115; Ellis et al. 2007: 472). Aldridge (2014) highlights the "inclusive and democratising aspects" (ibid.: 118) of participatory research approaches as participants are co-producers, co-analysts and disseminators within the research project (cf. ibid.: 124). Although a participatory approach does not diminish power relations entirely, as the researcher might for instance still decide whom to integrate and to invite to participate, it nevertheless increases awareness about asymmetric relations within the research process which gives participants a greater opportunity to challenge them (cf. Krause 2017: 20). Finally, integrating refugees and asylum seekers actively into the research process may also reduce the risk that the research reproduces a presentation of the respondents as passive victims and as a solely vulnerable group but instead contributes to enhancing their agency.

References

Aldridge, J. (2014) 'Working with vulnerable groups in social research: Dilemmas by default and design', *Qualitative Research*, 14: 112–130.

Baranik, L. E., Hurst, C. S. and Eby, L. T. (2018) 'The stigma of being a refugee: A mixed-method study of refugees' experiences of vocational stress', *Journal of Vocational Behavior*, 105: 116–130.

Beuving, J. and De Vries, G. (2015) 'Doing qualitative research', *The craft of naturalistic inquiry*. Amsterdam: Amsterdam University Press.

Bigo, D. (2002) 'Security and immigration: Toward a critique of the governmentality of unease', *Alternatives: Global, Local, Political*, 27: 63–92.

Binder, S. and Tošić, J. (2003) 'Flüchtlingsforschung. Sozialanthropologische Ansätze und genderspezifische Aspekte', *SWS-Rundschau*, 43 (4): 450–472.

Birman, D. (2006) 'Ethical issues in research with immigrants and refugees', in J. E. Trimble and C. B. Fisher (eds) *The Handbook of ethical research with ethnocultural populations & communities*. Thousand Oaks, London and New Delhi: Sage, pp. 155–177.

Brown, K., Ecclestone, K. and Emmel, N. (2017) 'The many faces of vulnerability', *Social Policy & Society*, 16 (3): 497–510.

Brunarska, Z. and Weinar, A. (2013) 'Asylum seekers, refugees and IDPs in the EaP countries: Recognition, social protection and integration – An overview', *CARIM-East Research Report, 2013/45*.

Bruns, B., Happ, D. and Beurkens, K. (2018) 'Risiken und Nebenwirkungen. Unbehagliche Begegnungen zwischen Forschenden und Beforschten', in F. Meyer, J. Miggelbrink and K. Beurskens (eds) *Ins Feld und zurück – Praktische Probleme qualitativer Forschung in der Sozialgeographie*. Berlin: Springer Spektrum, pp. 123–128.

Butler, J., Gambetti, Z. and Sabsay, L. (2016) 'Introduction', in J. Butler, Z. Gambetti and L. Sabsay (eds) *Vulnerability in resistance*. Durham and London: Duke University Press, pp. 1–11.

Chatty, D. (2014) 'Anthropology and forced migration', in E. Fiddian-Qasmiyeh, G. Loescher, K. Long and N. Sigona (eds) *The Oxford handbook of refugee and forced migration studies*. Oxford: Open University Press, pp. 1–9.

Clark, C. R. (2007) 'Understanding vulnerability: From categories to experiences of young Congolese people in Uganda', *Children & Society*, 21: 284–296.

Clark-Kazak, C. (2017) 'Ethical considerations: Research with people in situations of forced migration', *Refugee*, 33 (2): 11–17.

Council for International Organizations of Medical Sciences (CIOMS) and World Health Organization (WHO) (2016) 'International ethical guidelines for health-related research involving humans', Geneva, https://cioms.ch/shop/product/international-ethical-guidelines-for-health-related-research-involving-humans/ (accessed 19 October 2018).

de Wildt, R. (2016) 'Ethnographic research on the sex industry: The ambivalence of ethical guidelines', in D. Siegel and R. de Wildt (eds) *Ethical concerns in research on human trafficking*. Cham: Springer, pp. 51–69.

Donnelly, S., Raghallaigh, M. N. and Foreman, M. (2019) 'Reflections on the use of community based participatory research to affect social and political change: Examples from research with refugees and older people in Ireland', *European Journal of Social Work*, 22 (5): 831–844.

Düvell, F., Triandafyllidou, A. and Bastian, V. (2010) 'Ethical issues in irregular migration research in Europe', *Population, Space and Place*, 16: 227–239.

Ehrkamp, P. and Leitner, H. (2006) 'Guest editorial: Rethinking immigration and citizenship: New spaces of migrant transnationalism and belonging', *Environment and Planning A*, 38: 1591–1597.

Ellis, H. B., Kia-Keating, M., Aden, Y. S. and Lincoln, A. N. A. (2007) 'Ethical research in refugee communities and the use of community participatory methods', *Transcultural Psychiatry*, 44 (3): 459–481.

Emmel, N. and Hughes, K. (2010) "Recession, it's all the same to us son': The longitudinal experience (1999–2010) of deprivation', *Twenty-First Century Society*, 5 (2): 171–181.

European Commission (n.d.) *Guidance note – Research on refugees, asylum seekers & migrants*. Directorate-General for Research and Innovation, http://ec.europa.eu/research/participants/docs/h2020-funding-guide/cross-cutting-issues/ethics_en.htm (accessed 10 October 2019).

Freedman, J., Kivilcim, Z. and Baklacioğlu, N. Ö. (eds) (2017) 'Introduction. Gender, migration and exile', *A gendered approach to the Syrian refugee crisis*. Abingdon: Routledge, pp. 1–15.

Gibney, M. J. (2018) 'The ethics of refugees', *Philosophy Compass* 13 (10): n.p.

Gilmartin, M. (2008) 'Migration, identity and belonging', *Geography Compass*, 2 (6): 1837–1852.

Habib, N. (2018) 'Gender role changes and their impacts on Syrian women refugees in Berlin in light of the Syrian crisis', *WZB Discussion Paper*, SP VI: 2018–101.

Happ, D., Meyer, F., Miggelbrink, J. and Beurskens, K. (2018) '(Un-)informed consent? Regulating and managing fieldwork encounters in practice', in J. Wintzer (ed) *Sozialraum erforschen: Qualitative Methoden in der Geographie*. Berlin: Springer Spektrum, pp. 19–35.

Hovy, B. (2018) 'Registration – A sine qua non for refugee protection', in G. Hugo, M. J. Abbasi-Shavazi, E. P. Kraly (eds) *Demography of refugee and forced migration*. Cham: Springer, pp. 39–55.

Huysmans, J. (2006) *The politics of insecurity: Fear, migration and asylum in the EU*. Routledge: New York.

Hyndman, J. (2012) 'The geopolitics of migration and mobility', *Geopolitics*, 17: 243–255.

Jacobsen, K. and Landau, L. (2003) 'Researching refugees: some methodological and ethical considerations in social science and forced migration', *New Issues in Refugee Research*. UNHCR Working Paper No. 90, pp. 1–27. https://www.unhcr.org/research/working/3f13bb967/researching-refugees-methodological-ethical-considerations-social-science.html.

Kabranian-Melkonian, S. (2015) 'Ethical concerns with refugee research', *Journal of Human Behavior in the Social Environment*, 25 (7): 714–722.

Kaźmierkiewicz, P. (2011) 'Integration of migrants in Ukraine: Situation and needs assessment', http://www.osce.org/odihr/81760.

Keely, C. B. and Kraly, E. P. (2018) 'Concepts of refugee and forced migration. Considerations for demographic analysis', in G. Hugo, M. J. Abbasi-Shavazi and E. P. Kraly (eds) *Demography of refugee and forced migration*. Springer: Cham, pp. 21–37.

Kobayashi, A. (2005) 'Anti-racist feminism in geography: An agenda for social action', in L. Nelson and J. Seager (eds) *A companion to feminist geography*. Oxford: Blackwell Publishing, pp. 32–40.

Krause, U. (2017) 'Researching forced migration: Critical reflections on research ethics during fieldwork', *Refugee Studies Centre. Working Paper Series* (123). Oxford: Department of International Development, University of Oxford.

Lewis, H. (2016) 'Negotiating anonymity, informed consent and "illegality": Researching forced labour experiences among refugees and asylum seekers in the UK', in D. Siegel and R. de Wildt (eds) *Ethical concerns in research on human trafficking*. Cham: Springer, pp. 99–116.

Mackenzie, C., McDowell, C. and Pittaway, E. (2007) 'Beyond 'do no harm': The challenge of constructing ethical relationships in refugee research', *Journal of Refugee Studies*, 20 (2): 299–319.

Mackenzie, C., Rogers, W. and Dodds, S. (2014) 'Introduction: What is vulnerability, and why does it matter for moral theory?', in C. Mackenzie, W. Rogers and S. Dodds (eds) *Vulnerability. New essays in ethics and feminist philosophy*. Oxford: Oxford University Press, p. 1–29.

Maillet, P., Mountz, A. and Williams, K. (2016) 'Researching migration and enforcement in obscured places: practical, ethical and methodological challenges to fieldwork', *Social & Cultural Geography*. 18 (7): 927–950.

Makhoul, J., Chehab, R. F., Shaito, Z. and Sibai, A. M. (2018) 'A scoping review of reporting 'ethical research practices' in research conducted among refugees and war-affected populations in the Arab world', *BMC Medical Ethics*, 19: 36.

Markowski, S., Brunarska, Z. and Nestorowicz, J. (2014) 'Migration management in the EU's eastern neighbourhood', *Journal of Sociology*, 50 (1): 65–77.

Marlowe, J. (2018) *Belonging and transnational refugee settlement. Unsettling the everyday and the extraordinary*. London: Routledge.

McEwan, C. (2009) 'Postcolonialism/postcolonial geographies', in R. Kitchin and N. Thrift (eds) *International encyclopedia of human geography*, Vol. 8, Elsevier, pp. 327–333.

McLaughlin, R. H. and Alfaro-Velcamp, T. (2015) 'The vulnerability of immigrants in research: Enhancing protocol development and ethics review', *Journal of Academic Ethics*, 13: 27–43.

Migration Policy Centre (2013) *Migration profile Ukraine*, MPC.

Mountz, A., Miyares, I. M., Wright, R. and Bailey, A. (2003) 'Methodologically becoming: Power, knowledge and team research', *Gender, Place & Culture: A Journal of Feminist Geography*, 10 (1): 29–46.

Peoples, C. and Vaughan-Williams, N. (2010) *Critical security studies. An introduction*. Routledge: London.

Perry, K. H. (2011) 'Ethics, vulnerability, and speakers of other languages: How university IRBs (do not) speak to research involving refugee participants', *Qualitative Inquiry*, 17 (10): 899–912.

Peter, E. and Friedland, J. (2017) 'Recognizing risk and vulnerability in research ethics: Imagining the "what ifs?"', *Journal of Empirical Research on Human Research Ethics*, 12 (2): 107–116.

Pittaway, E., Bartolomei, L. and Hugman, R. (2010) '"Stop stealing our stories": The ethics of research with vulnerable groups', *Journal of Human Rights Practice*, 2 (2): 229–51.

Sabsay, L. (2016) 'Permeable bodies: Vulnerability, affective powers, hegemony', in J. Butler, Z. Gambetti, and L. Sabsay (ed) *Vulnerability in resistance*. Durham and London: Duke University Press, pp. 278–302.

Salter, M.B. (2013). 'The ethnographic turn: Introduction', in M. B. Salter and C. E. Mutlu (ed) *Research methods in critical security studies: An introduction*. New York: Routledge.

Shevel, O. (2011) *Migration, refugee policy, and state building in postcommunist Europe*. Cambridge: Cambridge University Press.

Sigona, N. (2014) 'The politics of refugee voices: Representations, narratives, and memories', in E. Fiddian Qasmiyeh, G. Loescher, K. Long, and N. Sigona (eds) *The Oxford handbook of refugee and forced migration studies*. Oxford: Oxford University Press.

Smith, K. and Waite, L. (2019) 'New and enduring narratives of vulnerability: Rethinking stories about the figure of the refugee', *Journal of Ethnic and Migration Studies*, 45 (13): 2289–2307.

UNHCR (2017) 'Ukraine refugees and asylum seekers', Thematic Factsheet, https://reliefweb.int/report/ukraine/unhcr-ukraine-refugees-and-asylum-seekers-june-2017 (accessed 10 October 2017).

UNHCR and IDC (2016) 'Vulnerability screening tool. Identifying and addressing vulnerability: A tool for asylum and migration systems', http://www.refworld.org/docid/57f21f6b4.html (accessed 19 October 2018).

United Nations (2016) 'New York declaration for refugees and migrants', 3 October 2016, A/RES/71/1, paragraph 52, www.refworld.org/docid/57ceb74a4.html (accessed 19 October 2018).

Vacchelli, E. (2018) 'Embodiment in qualitative research: Collage making with migrant, refugee and asylum seeking women', *Qualitative Research*, 18 (2): 171–190.

van Liempt, I. and Bilger, V. (2018) 'Methodological and ethical dilemmas in research among smuggled migrants', in R. Zapata-Barrero and E. Yalaz (eds) *Qualitative research in European migration studies*, IMISCOE Research Series, pp. 269–285.

von Benzon, N. and van Blerk, L. (2017) 'Research relationships and responsibilities: "Doing" research with "vulnerable" participants: Introduction to the special edition', *Social & Cultural Geography*, 18 (7): 895–905.

von Unger, H. (2018) 'Ethische Reflexivität in der Fluchtforschung. Erfahrungen aus einem soziologischen Lehrforschungsprojekt', *Forum Qualitative Sozialforschung/ Forum: Qualitative Social Research*, 19: 3.

6 Research ethics and inequalities of knowledge production in Eastern Europe and Eurasia

*Kristine Beurskens, Madlen Pilz and
Lela Rekhviashvili*

Introduction

Research ethics have gained enormous importance and visibility in recent academic debates. Humanities and social and spatial sciences have picked up on questions of ethics initiated in the medical and psychological sciences earlier. Critical debate on research ethics are proliferating in various academic strands and debates (e.g. Scheper-Hughes, 1995; Mullings, 1999; Streiner and Sidani, 2010; Bell, 2014; Belur, 2014; von Unger et al., 2015; Askins and Blazek, 2017). At the same time, concern has been raised in critical debates on research practices about Western/Northern dominance in global academic spheres, resulting in attempts to de-Westernise and decolonise research (e.g. Hann and Dunn, 1996; Tlostanova, 2015; Glück, 2016; Radcliffe, 2017). This paper sets out to contribute to such broader debates from the particular perspective of three researchers socialised, educated and currently employed in a, broadly conceived, Western academic context, with personal experiences of living under socialism or in a post-socialist context, and researching Eastern Europe and post-socialist Eurasia throughout the past decade. Our contribution asks what specific ethical challenges are at play when researchers trained in the Western academic world do fieldwork in post-socialist Eastern Europe and Eurasia or, more generally, how power asymmetries and different traditions in the production of knowledge affect ethical challenges in qualitative fieldwork.

The starting point of this inquiry is a long ago stated but continuously relevant concern of marginality in research about Eastern Europe and Eurasia (Hörschelmann, 2002; Suyarkulova, 2018). Since the early 2000s, scholars have emphasised a problematic academic job division in which Eastern Europe has served as an empirical testing ground for Western, primarily Anglo-Saxon theories (Timár, 2004), while theories produced in the region were often seen "as limited, parochial and only local" (Stenning and Hörschelmann, 2008, p. 315). Thus, almost three decades since the collapse of socialist regimes, Eastern Europe and even more so of peripheral Eurasia have contributed marginally to theory production (Tuvikene, 2016; Gentile, 2018; Trubina, 2018) and have often been left out of comparative research

DOI: 10.4324/9780429507366-6

agendas both in the Global North and South (Müller, 2020). Yet another concern is that benchmarking of the post-socialist region against Western norms reproduces othering of the region in terms of its, so-called, relative backwardness (Stenning and Hörschelmann, 2008; compare also Hann and Dunn, 1996; Verdery, 1996; Brandtstädter, 2007; Berdahl, 2010; Thelen, 2011; Wiest, 2012). Departing from acknowledgements of uneven geographies of knowledge production, we are further inspired by recent post-Soviet decolonial thinking on global imbalances in knowledge production (Tlostanova, 2015) and our own observations of frictions between Western academic ethical standards and our experiences of ethical challenges in Eastern Europe/ Eurasia. We discuss implications of such coloniality of knowledge production primarily for research ethics and researchers' responsibilities in the field, and also ask how, if at all, can we contribute to de-Westernisation and decolonisation of knowledge production, while remaining dependent on, but also privileged through, Western academic structures and resources.

The chapter reflects on the ethical challenges that play out when researchers trained in the West do fieldwork in post-socialist Eastern Europe and Eurasia through the lens of our own fieldwork experiences. We take into account our different individual backgrounds and belongings (USSR, GDR, Georgia) and our academic training in different disciplines (Geography, Social Anthropology and Political Sciences) at Western universities (CEU Budapest, HU Berlin). In the following three sections, we discuss how our positionality together with a general hegemony of the West in knowledge production shaped our fieldwork experiences, which ethical guidelines we applied and how those guidelines helped or constrained us in addressing ethical challenges in the field. Motivated by the question of how we could have done better, we stay in dialogue with decolonial literature, alongside ethnographic self-reflexivity on specific positionalities in these contexts (Steger, 2004; Roth, 2005) and the possibilities of redressing power imbalances through participatory approaches (Lassiter, 2005), allowing us to acknowledge the coloniality of the knowledge production processes we have been engaged in. We reflect on the situatedness of ourselves as researchers and participants in our respective research projects also within a global field of power relations that equally shape our emerging research relations. Inspired by the key representative of post-Soviet decolonial thinking, Madina Tlostanova, we concur that "only after people realise their own coloniality of thinking and of being (...) that they can start elaborating their own theory" (Tlostanova, 2015, p. 280).

The chapter contributes to but also moves beyond existing discussions on doing qualitative research in Eastern Europe and Eurasia. The multidisciplinary literature on research methods and ethics in the region stresses (1) the effects of constant changes on everyday lives, (2) the legacies or continuities of authoritarian politics and (3) the distance between former Eastern and Western Bloc experiences, which shape relations during fieldwork (Burawoy and Verdery, 1999; Humphrey, 2002; Herzfeld, 2010). Many publications

address the speciality of Eastern European states in the sense of their conditions as *transformation societies* or their present or former *authoritarian*
regimes (e.g. Dudwick et al, 2003; Michailova and Clark, 2004; Schäuble
et al., 2006; Roberts, 2012; Glasius et al., 2018). As Soulsby (2004, p. 43) puts
it, "(t)he pace of change in these societies offers an exceptional and exciting
challenge," giving not only specific insights into societal conditions, such
as the newly emerging struggles for welfare under market conditions but
also deliberating aspects of democratic rules, new freedoms and possibilities. Studying societies in such a transition period also demands sensitive
and flexible behaviour from researchers. Such an argumentation is also supported by Michailova and Clark (2004), who observe that doing research
in transformation societies, due to constant processes of change, can make
researchers face particular difficulties and heightened insecurities during
the fieldwork. Another strand of academic debate in this context discusses
the ethical position of the researcher. Steger (2004), for instance, describes
his personal ambiguities as a researcher from the outside, not having personally experienced everyday life in a socialist society. De Soto (2000) similarly
discusses how her high *a priori* personal demands of doing feminist, contextual, inclusive and socially relevant research were confronted with unexpected attributions and suspicion when entering the post-socialist research
field as a Western researcher. She describes how her East German research
participants' feelings of being dominated by West Germans impacted their
interactions with her as a Western researcher as well.

While discussions and calls for decolonisation and de-Westernisation
of research methods and ethics have become prominent among scientists
from the Global South and in the Western academic world (Abu-Lughod,
1975; Chakrabarty, 2000; Mignolo, 2002; Escobar, 2007; Robinson, 2011;
Glück, 2015), post-socialist Eastern Europe and, even more so, post-Soviet
Eurasia remain largely outside the discussion. The region has been "left out
once again from modernity/coloniality[1] and dewesternisation and deco-
loniality alike" (Tlostanova, 2015, p. 268). The absence of post-socialist
Eurasia from decolonial perspectives and the difficulty of elaborating dialogues with post-colonial thinking, despite numerous calls for post-socialist
post-colonial dialogues since the 1990s, is related to the invisibility of the
coloniality and racialisation of power throughout socialist times, especially
within the Soviet Union (Tlostanova, 2018). In other words, the challenge
is not only that Eastern Europe and Eurasia are marginal to knowledge
production spaces in general, as numerously articulated in critical scholarship on the post-socialist region (Timár, 2004; Hörschelmann and
Stenning, 2008; Müller, 2020), but also that the region is often excluded from

---

1   Decolonial thinkers define coloniality as the process of reproduction of colonial forms of
    domination after the end of colonial administration in the realms of power, knowledge
    and being (Maldonado-Torres, 2004; Escobar, 2007).

counter-hegemonic knowledge production streams, such as decolonial theorising itself. Taking the knowledge asymmetries seriously, the particular contribution of this paper is to consider how such knowledge asymmetries affect qualitative research ethics in Eastern Europe and Eurasia.

We show in our contribution how asymmetric power relations within the production of knowledge lead to grave ethical challenges. The problem for us here is not so much that Western ethical guidelines, academic discussions and reflections of ethics fail to fully prepare us for concrete challenges, as this is unsurprising and probably holds true for many research contexts. Similarly, it is a given fact that East European and Eurasian research contexts display specificities which researchers have to consider and reflect on. The problem for us is rather the assumption of the universality of Eurocentric knowledge and values. Our experiences reveal that the key underlying assumption of Western ethical thinking – that academic knowledge accumulation is valuable in its own right – can be fundamentally questioned by research participants who are struggling to make ends meet on a day-to-day basis, aware that knowledge can be turned against them, and ask for tangible outcomes and benefits of our research for their daily lives. Our narratives illustrate that our lack of preparedness for ethical challenges was linked primarily with our underestimation of how deep power differentials can be between us and our research participants and even of the limits of our cultural and political embeddedness in respective locations. Most importantly, we show how and why our limited capacity to benefit research participants, to produce locally relevant knowledge and to share knowledge with local audiences is possibly the most pressing ethical challenge. Our concerns then resonate with concerns of colleagues working in the region, as Mohira Suyarkulova summarises in her intervention into the debate on research ethics in Central Asia on the media platform openDemocracy:

> There seems to be a sense of entitlement on the part of the researcher to have access to the field in order to know 'the truth'. While it is laudable to pursue the truth concealed by power, one ought to question their own motivations: why do you need to know? Who is this scholarship for? Whose needs does it serve? In the case of Central Asia, all too often scholarship serves outside audiences, with most of the findings published in a foreign language in obscure academic journals hidden behind a paywall, thus making them virtually inaccessible to the region's citizens.
> (Suyarkulova, 2018)

In the following three sections, we discuss our personal experiences of facing ethical challenges in the field in Eastern Europe and Eurasia. Summarising our personal accounts, we will then conclude by sharing our imagination of collective and collaborative tactics to contribute to subverting Eurocentric knowledge production systems and, thus, challenge existing knowledge production geographies.

Power asymmetries: experiences in the research field and in academia (Madlen Pilz)

I carried out my ethnographic research in Georgia between 2008 and 2012, being listed as PhD faculty staff at the Institute for European Ethnology of the Humboldt University in Berlin, Germany. In what follows, I would like to share two experiences that unsettled my initial expectations about ethics in fieldwork. The first one touches on my interpersonal relations with research participants in Georgia. In my discussion of this example, I will outline how our relations were shaped by global power hierarchies and what kind of impact they had on my fieldwork. The second experience reflects the difficulties I encountered when seeking to make sense of my empirical data back home against the background of concepts formulated to describe Western European realities. The first aspect taught me about my embeddedness in Western life-worlds, the second one made me aware of the dominance of Western academic knowledge production.

In retrospect, I must admit that I entered the research 'field' with a major misunderstanding about myself and the ways I would be perceived in Georgia. Having grown up with a Soviet passport in the 1980s and because of my recent activities in different left-wing political initiatives after 1989, I did not feel like a representative of the 'West' when I arrived for research. Despite being an absolute newcomer to Georgia, I was convinced I would gain people's confidence on the ground of our shared socialist past. But people in Georgia welcomed and confronted me with a different perspective. They saw me as someone from the 'West,' i.e. as someone from that part of the world where all the political advisers came from who, together with the Georgian government, were producing social devastation in the country. In their minds, I was from that part of the world where people lived a better life, where they had possibilities to develop their careers, to advance personally and economically. Indeed, their perception of the asymmetry between the two countries – grounded in the globally uneven settings of economic development and political power relations – was more than true. Life in Georgia for the majority was very much about survival, grounded in familial networks and in a soft blending of formal and informal economic practices. Due to this situation, for many, the initial perception of me was guided by their necessity of making ends meet, finding economic resources, business possibilities, etc. Consequently, I was imagined as a chance, a gate-opener, someone who could provide access to the 'West,' whether it was for studying there or to mount a business venture – expectations I was unable to meet. Hence, the global asymmetries caused sharp imbalances between people in Georgia and me. Our mutual expectations of each other were influenced by two key aspects: firstly, sharp economic differences, as I received pay for conducting my PhD research for a 'Western' academy, while the people I met always faced the necessity of improving their incomes. Secondly, our divergent possibilities to practise a sense of reciprocity in different kinds of personal relations. Even if I could not meet their hopes and desires, many of them helped me to achieve my aims: they acted as gate-openers for me into the local social context and provided me with the insider knowledge that I needed to deal with sensitive situations. By contrast, I seemed to be as 'useless' to them as a beautiful shard, a flashing eye, and, thus, I confirmed rather than confounded stereotypes about the 'West.'

In order to tackle these asymmetries, my research partners and I developed different but almost symbolical strategies. Some, for example, asked me to use their services at slightly higher rates than usual in the local context. For others, our relations could be used as a type of social capital within respective networks. I was also asked for advice in different situations and, at times, a kind of friendship developed. I generally tried to compensate the imbalance of reciprocity by deploying the social and cultural capital attached to my position as a researcher, as I had no serious economic resources to offer. When asked about the 'West,' I tried to provide research participants with detailed knowledge and critical insights – according to my political position and academic knowledge – about Germany and the variety of (legal) avenues for entering the country.

Another factor which helped me, although it was not a particular merit, was time. As the project aimed to observe post-socialist change over a *longue durée*, I was able to organise my ethnographic research over a period of four years, which I used for visiting Georgia frequently. Thus, I had the chance to accompany my research partners for several years of their lives. During my stays in Georgia, I tried to share their living conditions as much as possible, and in between my travels, I stayed in touch via e-mail and Facebook. We lived, therefore, through different situations together, and this common experience gave rise to many unintended talks about everyday life. I tried to ensure that the privilege of sharing in participants' everyday life in Georgia was matched by me also sharing as much information as possible about my own life in Germany. Furthermore, being sensitive to the complex norms, expectations and hierarchies that can be involved in gift-giving, I sought to select gifts for participants carefully, selecting items from Germany that corresponded with the interests of my partners. I also invited them for German dinners, which I used to get them into contact with each other, and I offered to host them if they wanted to visit Germany. Over time, people started to recognise my growing knowledge about the social and political situation in Georgia, and they were very interested in listening to my experiences and reflections about the processes that I was witnessing. Additionally, my insider knowledge of former Soviet everyday life helped to establish commonalities between my research partners and me, as long as we shared similar ways of reflecting on that period. We increasingly evaluated my empirical work and emerging analytical themes together through conversations in which I shared my observations and reflections. As I knew that nearly none of my partners would be able to read my PhD in German, I saw these 'collaborative moments' (Lassiter, 2005), in which we generated new knowledge together, as one possibility to give back some of my knowledge and to open up my findings to critical reflection by participants.

Analysing the effects of globally produced asymmetries on research relations is part of daily scientific practice in decolonial thinking and an inherent part of our professional self-understanding. The ethical implication of this is that each investigator has a responsibility to reflect on the kinds of asymmetries that are at play in research situations and to be aware of her or his potential role in reproducing or countering and subverting them, as, for example, I did by building up alliances with people in the 'field.' This means to 'invest time' in commonalities that allow the construction of shared symbolical spaces which can foster proximity and trust, something that has long been acknowledged as a core requirement of anthropological research (Lassiter, 2005). Globally produced asymmetries, as a form of coloniality, are enhanced by the monopoly of knowledge production over 'others'/'other' territories. By contrast, sharing knowledge or the common generation of knowledge in 'collaboration' can, thus, be seen as a tool to disrupt and subvert colonial logics and mechanisms. This collaborative knowledge production should further be addressed properly in the final output. In my case, I would have to acknowledge that I did not do this well enough in my PhD thesis. In hindsight, I recognise that I would have needed to think about and experiment more thoroughly with new ways of description.

As my fieldwork example illustrates, time is a significant prerequisite regarding the points discussed, allowing the building up of commonalities, sharing of knowledge and representation of the collaboration in the final output. Due to growing logics of economisation, time is, however, increasingly scarce in Western academia. In my case, the 'unusually' long period that was planned into my research allowed me to make at least a partial personal contribution. However, this was not always valued positively in my academic context. Consequently, I argue that structural conditions play an important role in facilitating or obstructing researchers' potentials to think of and practise ethical research relationships and to properly recognise research partners' contribution to academic knowledge production.

The second aspect of my research that I wish to discuss is the post-fieldwork analytical period back home. During my research, I observed the protests taking place in Tbilisi from April to June 2009 with tens of thousands of people on the streets each day. Some of them spent months at the four protest camps in the city, thus, paralysing the entire infrastructure of the city. I collected as much information about the protests and how it related to people's everyday lives as I could. Back home, I aimed to analyse post-socialist transformation in Georgia as a politics of exclusion generating real social effects on people's everyday lives. The strategic use of discursive classifications, such as 'Soviet,' 'Georgian' and 'Western,' was central to the new social order when they were referred to people's different educational backgrounds and described social processes, such as the declining value of the soviet education and knowledge of the Russian language on the labour market in comparison to Western education and knowledge of the English language (Mühlfried, 2006; Jones, 2013).[2] Hence, I saw the protests as a magnifying glass allowing me to analyse the connections between post-socialist discursive strategies and numerous problems in people's everyday lives.

When reviewing the scientific literature and media coverage of this and former protests in Georgia, I found myself confronted with an epistemological inconsistency. Instead of recognising the Georgian protestors as political actors and their critical arguments as political, the existing analyses were speaking about them as people seduced by the political opposition or driven by a persistent (corrupt) Soviet habitus which was the reason for their unemployment and, subsequently, for their protest. Consequently, the scientific discourse represented Georgia as a weak or nearly inexistent civil society (e.g. Kaufmann, 2009; Nodia, 2013). Others (Cheterian, 2013) analysed the regularly renewed protests in Georgia as uprisings of the urban educated middle classes, which he considered to be structured around various branches of civil society organisations (NGOs) but lacking the pressure of a social movement.

At some point, I realised that the problem was grounded in their classical notion of civil society, describing it as the sphere of collective action which is structurally situated between the state and the family, holding the government accountable and assuming public tasks supporting society's cohesion (Nohlen, 2002), a sphere that is equated with NGOs in all works about Georgia (e.g. Cheterian, 2013; Kaufmann, 2009). For the most part, the protestors and their actions were, at least in 2009, not graspable by this definition. As a result, they were not defined by scholars as part of civil society or as forming a social movement. The protests of 'normal' people on the streets could not, therefore, be interpreted as intelligible political action. Instead, they were transformed into an event that was arbitrarily interpreted as unpolitical.

My observations here connect with the conceptual discussion on decolonial thinking as follows: firstly, the hegemony of Western concepts such as civil society, that were developed in Western contexts but later universalised as a blueprint for democracy, produced the invisibility of other explanations and ways to organise political agency (Hann and Dunn, 1996, p. 7). Secondly, an epistemological bias with serious ethical implications results: the lack of scientific recognition of the actors and their protest which, in the Georgian context, fed into local politicians' defamatory rhetoric about the protests and the protestors. A social science that reduces itself to legitimise exclusionary politics is, in my point of view, provoking serious ethical

---

2   The 'Soviet' was generally used to describe an inappropriate kind of habitus that corresponded very often with everything undemocratic and corrupt, meanwhile a 'Western' habitus was readily equated with being democratic and modern, and the 'Georgian' could stand for incorruptness as well as for virginity and a certain fighting spirit (Pilz, 2019).

questions about its practices, objectives and positioning in society. Thirdly, this point connects to another epistemological bias: as unrecognised political actors, people become defined as not yet rational and active enough. This also implies that they are not yet developed and democratic enough, i.e. not yet sufficiently Western. Social and political research here runs the risk of reaffirming global colonial hierarchies in the relations between post-socialist space and 'the West' (Tlostanova, 2015). The epistemological border in this setting can be seen as twofold. On the one hand, it seems to be one of the research methodologies in the field, i.e. the distinction between deductive and inductive methods. On the other hand, taking a closer look at the political positions and intellectual contexts of many theorists and researchers, it is also a specifically Western perspective on the East that is trapped in conventional modes of Western knowledge production, even when they are applied by Eastern scholars.

Against this background of political rhetoric and scientific analysis, I had to find an analytical approach that allowed me to include interpretations of all of the information that I had gathered about the protest and find notions that were suited to explaining the local particularities of the protest as well as to point to the universal features it shared with movements elsewhere. In order to do so, I turned to literature about urban protests in Europe and Latin America, compared approaches and notions with those used in the post-socialist literature, and started to focus centrally on aspects of practice and agency in civil society. Studies on citizenship in Latin Amerika (Holston, 2008) were particularly helpful here as was the heuristic concept of civil society that Michał Buchowksi (1996) formulated in his analysis of how protests were organised in socialist and post-socialist Poland. These publications were reconsidering the specificities of political settings and resulting path dependencies. They helped me to recognise different forms of organisation as political action and to redefine the social question as a political one.

Navigating group discussions in Russian Karelia (Kristine Beurskens)

My education as a researcher was based in a Western academic context, studying in Germany, the Netherlands and the United States and gaining my PhD and first research experiences in Germany. However, I grew up and experienced life in a socialist country, i.e. the GDR, until the age of 14. From 1989 onwards, I then lived in a country that, step by step, adopted Western institutional structures, including the education system. Doing research in a post-socialist context made me aware of my paradoxical position. My personal background would regularly put me in situations where, in the middle of some fieldwork interaction, I would suddenly feel very sensitive and aware of stereotypes, assumptions and aversion that were 'in the air.' These felt stereotypes would range from assumptions about the presumed backwardness of 'the East,' fears of and aversions against Western dominance, and assumptions about possible secret wishes and demands regarding the interaction. Such stereotypes generally forwarded certain understandings of power relations in this interaction and conveyed levels of discomfort. At the start of my career, such feelings, as all rather personal sensitivities in the field, seemed to conflict with my planned schedule and approach. However, it has since been more widely recognised that such emotions are not uncommon in social research. Emerging debates on emotional aspects of our research positionalities and practices are now supporting a more differentiated and sensitive approach to such matters (e.g. Bondi, 2005; Askins and Blazek, 2017). Combined with a heightened sensibility for decolonising and de-Westernising our research, this means to me that we should pay attention to anything that helps us to be more sensitive and aware of diverse issues, sorrows and feelings in the field. I see this as helpful for developing more ethical research practices in East European and other post-socialist contexts.

Most of all, it encourages me to enter into a stronger dialogue on how we do research, with whom and, most of all, for whose benefit. As qualitative research in its practice comes with a number of ethical implications regarding the involved relation between researcher and the researched, there is one thing we can pay special attention to (Ries, 2000). We can raise the attention of the people we meet in our research (anywhere) to their individual and collective agencies and the possibilities of making use of these chances, reveal the ways of how to create impact on structures, but also unravel colonist aspects, even the ones we produce or reproduce ourselves.

One particular experience when such issues of power relations in a setting of interacting science cultures between East and West came to the fore was my research in Russia in 2008 and 2009. Together with another young female German colleague, I was studying border relations between Finland and Russia in a rural area of Karelia, a few hundred kilometres north of St. Petersburg. We were supported by two Russian researchers who had both been based at a Geography Department at a university on the Finnish side of the border for a couple of years. When I think back, I can recognise some profound instances of clashes of science cultures and accompanying ethical questions.

One such instance relates to the conceptualisation of the methodological approach. My research was part of a larger project analysing border relations at four different Eastern European border locations. Since our research was focussed on petty traders at the borders who were working informally, we wanted to find a method that was not too confrontational. We assumed that a shared, informal knowledge could best be found in group discussions. Reflecting back on this assumption, I depict a confrontation of science cultures and dominance of Western research standards over an Eastern European context: the concept of the methodological approach took place in our institutional context of Western academia, in a Western research institute and among researchers trained in Western conceptualisations and research approaches. Some members of our team brought fieldwork experiences from Eastern Europe to the discussions, but our choice was mostly guided by the preference for then-newly popular and favoured methods in qualitative social research in Western Europe. The decisions in favour of group discussions were made mainly because this method, following the principles of Western thinkers such as Karl Mannheim, promised to find out about the implicit knowledge of discussants; the underlying but not explicitly mentioned orientations of their actions (see Garfinkel, 1967; Mannheim, 1980; Bohnsack et al., 2010). However, qualitative research methods such as these that are based on the interpretive paradigm and that have been developed to closely observe and understand people's actions come with some cultural and ethical implications. According to some Western ambassadors of qualitative social research, the act of interpretation or the analysis of meaning has become a necessary tool for everyone in everyday life in an increasingly complex world, where roles and identities are not as determined as they were previously. "The openness of interpretation creates possibilities for communication within a global society in which the most diverse positions come into dialogue" (Knoblauch, 2008, p. 230, author's translation). This brief quote from the backgrounds of qualitative research methodology illustrates the conditions under which such methodologies have been developed. While an increase in openness and dialogue between diverse positions might indeed have been characteristic of social developments in Western societies since the 1960s, it cannot be taken for granted in every part of the world. In societies with authoritarian leaderships, for instance, in the light of the arbitrariness of its regulative powers, it is hard to know what one can say or do without disturbing consequences (Glasius et al., 2018).

This is the kind of issue that we were facing in our research in 2008. The mismatch in the very basic foundation of the approach of this research had prevailing impacts throughout the whole process of this study. One memorable event was a workshop to prepare the group discussions (Bruns and Zichner, 2009). Back in 2007/2008, group discussions were not yet a very widespread method in our spheres of spatially related research, and we assumed that our East European research partners would also not be very familiar with it. We organised a workshop led by two well-known German scientists in the field of qualitative methods to introduce us and our six invited colleagues from several countries on the Eastern border of the European Union to the method. Reflecting on this step in terms of science culture, this was, in some ways, a positive measure: we established knowledge collaboratively which would then get into an exchange with other academic contexts. However, our main intention was a different one, namely, to smoothen the path for our investigation. We hoped that, after the workshop, our collaborators would be able to explain the method and its key principles to potential research participants and to gate-keepers in the field. Most importantly, we hoped that the collaborating researchers would be in a better position to moderate the group discussions in a way that matched our approach to qualitative research.

Our workshop discussions quickly led us to understand, however, that it would be difficult to establish a common understanding of our approach in practice. We, therefore, decided that the only structuring element of all the planned group discussions should be a common opening question on recent changes of the border regime. At the workshop, we also brought up the issue of where to meet for the group discussions, as the protection of participants was an important issue to us. Because the people we planned to interview were partly involved in illegal cross-border activities, they would not have wanted to be seen talking to (Western) strangers in public. Following the suggestion of some collaborators, we agreed on somewhat hidden meeting places, possibly in the backrooms of cafes and restaurants. We also discussed the possible effects of the planned presence of us as researchers at the group discussions and had quite a long argument on adequate incentives for participants. During these debates, it became obvious how difficult it was to keep a constructive balance between our considerations and aims and the understandings, doubts and objections of our research partners. Some of the invited Eastern European scholars were more and some were less open to the whole idea of the approach. In the weeks following the workshop, we pondered some thoughts and questions. Did the workshop leave enough space for their doubts and possible interventions? Would a workshop carried out in an Eastern European location have supported a more balanced sphere, more geared towards the specific requirements of the research locations and respective local researchers, who possibly knew best what does and does not work in their research contexts?

The actual application of our well-prepared (so we assumed) qualitative research on groups in Russia was another instance of opposing cultures of science. We had planned the implementation in detail: a rather informal, inconspicuous place for the discussion, incentives for participants and a convivial atmosphere including the serving of snacks and drinks. Our understanding of sound comparative research at this point led us to aim for settings that were approximately the same in each of our fieldwork locations along the eastern EU border. But this turned out very differently. To our surprise, our Russian collaborators had chosen a very public meeting place for the group discussion, a conference room in the city hall. The room was decorated with flags and emanated formality. There were no snacks or drinks to establish a welcoming atmosphere. At this moment, in the field, we were perplexed, but there was no way around it. We had to take the set-up as it was and felt very dependent on our collaborators, who, by then, clearly had taken over the power of handling this aspect of the research.

Where did it go wrong? Did it 'go wrong' at all or was this simply a natural adaptation of our approach to the field? Should we have taken more time and engaged more openly from the start in order to be able to sense even small degrees of discrepancy between our methodological understandings and those of our collaborators and discuss their doubts and ideas with more interest and tolerance? Should we have taken a step back from our fixed common schedule for all the different locations and given the local researchers more responsibility? A point that strikes me when looking back is that our well-meant ethical intention of keeping participation in these discussions 'invisible' to the eyes of formal institutions had been ignored in the Russian case. Perhaps one of the reasons behind this is an incommensurable set of perceptions of (il)legality and (il)legitimacy in different parts of the world. In a research setting where people no longer believe in state actions and the state officials themselves are the most powerful pilots of the skies of illegality (Müller, 2013), being seen in a rare meeting in a small town administration could very well just be part of a game that we do not understand. Learning from such experiences also means being prepared for another type of interaction between Western academia and Eastern research contexts: to leave presumptions behind, question the foundations of our intentions and approaches, meet at eye level and be open to collaborative ways of conducting field research, such as to ensure sensitivity to local conditions, and benefit (and at least not harm) all of the parties involved.

Outsider in the native context: on the challenges of benefiting research participants (Lela Rekhviashhvili)

My career profile is specific but certainly not unique. I was socialised in the Western academic context of the Central European University (CEU, Budapest) but have done most of my empirical research in my home country, Georgia. Throughout my relatively short academic career, I have relied predominantly on qualitative methods and studied informal economic practices of vulnerable and marginalised groups, such as informal petty traders, informal parking guards and transport workers. In addition to my research in Georgia, I have also conducted fieldwork in Kyrgyzstan, examining drivers' working conditions for a period of two months. Despite different degrees of cultural closeness to my research subjects, in both cases, I entered the field as a researcher largely educated and socialised in Western academia, informed primarily by theories, methods and ethical standards that had little to do with knowledge production processes in the localities I was studying.

  While I do not think any ethical guidelines can prepare us completely for each context, neither can our so-called insider knowledge. Let me give an example. When I first returned to Georgia for my first ethnographic fieldwork there, I had been out of the country for two years in order to study in Hungary. I had visited regularly during that time and stayed closely tuned to political events in Georgia. However, neither my broad knowledge of the political situation nor what I saw as fairly good training in research ethics could allow me to fully anticipate the ethical challenges I would face during my research. Back in 2012, I intended to conduct fieldwork with Tbilisian street vendors, but it was extremely difficult to approach vendors and achieve what one could call informed consent for their research participation. This was predictable. As street vending was illegal, I expected that vendors were sensitive about talking to a stranger about their working environment. However, I only learnt over time, after spending weeks in my fieldwork locations, that there were more specific reasons for the vendors' mistrust. Firstly, the state-enforced violence towards the vendors was more severe than I first envisioned. Secondly, vendors were particularly scared during the electoral period, which was approaching in the summer of 2012, as they knew – and as I learned from them later – that the Georgian government often changed their treatment of street vending in electoral periods. In some cases, the government would ease their control of street vending to gain vendors' support. In other cases, however, the government would vigorously 'cleanse' the streets of Tbilisi from vendors to appeal to lower-middle- and middle-class electorates. Hence, I only learnt about the extent of my first and most important ethical challenge – to inform research participants of my research and have their consent to participate in the research – through doing preliminary research. I had to be very careful not to incur additional trauma by even simply activating fears of being a government-sent (or even an oppositional) undercover agent. By becoming able to name and understand some of the particular sources of vendors' insecurity and fears, I was better able to address the latter upfront with each new research participant. I attempted to reassure them through extensive discussions about the purposes of my research and my position as a student, while also learning how to share my personal political positions more openly, sincerely distancing myself from both the government and opposition forces. It took not only sincerity but also time for the vendors to observe me, to see how I communicated in front of them with police or city hall officials, to make their own judgements about my loyalties or autonomy.

Reflecting on my research experiences in Georgia, I would argue that both general ethical guidelines and prior knowledge of the context are important but that none of them can fully spare us from facing unforeseen ethical dilemmas. In this sense, while broadly Western perspectives on ethical research might indeed not prepare us for specific challenges of doing research in different parts of the globe, I would not necessarily judge the ethical guidelines on that ground. Ultimately, when specific aspects of such guidelines come into conflict with needs on the ground, scholars are given an opportunity and find ways to justify their ethical choices. Western academic debates on research ethics have for decades reflected critically on power relations and researchers' positioning in the field, asking who speaks for whom, whose agency and voice are considered, and criticising researchers' claims to objectivity. Nevertheless, I came to see and experience a few interrelated limits to my preparedness for ethical challenges in the field: the first challenge is the assumption of being entitled to study the lives of others in the name of contributing to academic knowledge production. A second closely linked issue is the question of benefiting research participants. And third is the question of working primarily with Western concepts, which can mean that we are sometimes unable to address locally pressing socio-political challenges that may, in turn, hamper our ability to benefit participants in a meaningful way.

Our ethical standards at the Central European University were set by the Ethical Research Policy, a document that was heavily rewritten and relatively sensible to the East European context but still largely relying on the Warwick University Guidelines on Ethical Practice. It, hence, set out a few principles that resemble other Anglo-American ethics guidelines, including doing no harm, ensuring participants' consent and continuously reflecting on who benefits and/or bears risk from the research. Framed in this way, ethical standards do not question a researcher's right to observe, interpret, analyse and narrate others' lives. Post-colonial critiques and feminist scholarship have, however, drawn significant attention to this aspect of power in knowledge production (Askins, 2018). For me, it is a key ethical question to consider. The requirement to ensure participants' consent acknowledges this in some way. However, I believe that the issue goes deeper. It is not simply about convincing research participants to give informed consent but also about reflecting critically on researchers' rights and entitlements of conducting such research in the first place. While we may not be asked to consider this more fundamental question in our institutional ethics procedures, research participants often, and validly, raise it, as 'not being harmed' and 'contributing to science' are not sufficient reasons to participate in research. What they are concerned about is the capacity of our research to actually transform and even improve their lives.

Indeed, one of the clearest ethical challenges I have faced that relates to eurocentrism and coloniality of knowledge production is the question of benefiting research participants. My fieldwork participants have continuously challenged me with the question of benefiting the communities I am studying. The absolute majority of *marshrutka* drivers I have talked to have been extremely surprised about my intention to study their working conditions, while some representatives of *marshrutka* operating companies were explicit in drawing boundaries of what is 'not to be learned' by a researcher about their business. Street vendors, for example, would ask very explicitly how I could help them out, how could they benefit from my research, in our initial encounters. I usually responded by saying that they personally would not benefit from my research, but their help would support knowledge production processes that might, or might not, at one point in time be reflected in civil mobilising or policy-making. Importantly, it was not clear whether such a 'reflection' would translate into anything beneficial for the vendors. I believe it would be unfair to stir up any unrealistic expectations and I was often lucky to have generous research participants, willing to spend their time supporting my research. However, the challenge of not giving back remains severe. Not only would many of my research participants not be able to read research outcomes written in English but also the little contribution that my research could possibly make to knowledge production would hardly affect respective societies (let alone the research participants themselves).

This challenge seemed less pressing for me while I was doing my PhD research in Georgia. I did find time to write articles for diverse local media portals, be present for public discussions, hold presentations at universities in Tbilisi and collaborate with local activists. It would obviously be an exaggeration to say that I have really benefited my research participants in any direct way, but at least I attempted to share my research participants' concerns and perspectives and contribute to public discussion on the difficult working conditions of street vendors or public transport drivers. I saw that while different governments used similar strategies of repressing already marginalised communities, local civil society actors got better equipped to argue and mobilise against criminalising and illegalising informal workers over time. It was much harder to contribute to any public debates or to at least make my research accessible in my second fieldwork location, Bishkek, in 2016. Having spent only a relatively brief period of time there, without the ability to visit the city again easily, and having far less access to local media and activist circles, I see how useless my research can appear to be for both my research participants there in particular and Kyrgyz academics and citizens in general.

New ways have, of course, been found by ethically aware Western researchers and institutions to 'give something back' to communities whose resources are spent on knowledge production, be it through encouragement of open access publishing, or insisting on policy-relevant outcome formulations, impact evaluations and disseminating research findings through diverse, non-academic media. Going further than such often superficial measures, critical researchers in different social science disciplines abandon researchers' objectivity claims and engage consistently in processes and practices that seek to advance the causes of the communities they are studying. However, none of these measures alter the structure of knowledge production fundamentally where we as researchers are judged based primarily on our capacity to engage with transnational discourses, namely via English language academic spaces, publish in high-ranking journals or with prestigious publishing houses, and contribute to knowledge circulation primarily in the Global North. This not only affects our ability to ensure research results reach respective communities, in my case in Georgia or Kyrgyzstan. It also implies that the research questions we work with, the concepts we rely on and the types of argument that we frame are all moulded to fit the communication style and norms that current transnational academic space expects. This not always but quite often implies that we address questions and problems that are marginal to respective communities, and we also miss addressing those that are pressing, politically relevant, contested and locally in need of examination.

This discussion then leads me to my final concern: the limits that are created by our reliance on Western concepts for making our research politically relevant in the local contexts that we study. I have to admit that, prior to embarking on my first important field visit as a doctoral student examining street vending, I had not given proper thought to how and why my extensive reliance on Western concepts could become a challenge. In turn, having grown up in Georgia, I also did not anticipate that the local fieldwork context might be unfamiliar or require a particular adaptation from me. Unsurprisingly, I was proven wrong on both accounts. What I ended up finding in the field was that I had to constantly 'translate' my research. Such a translation involved actual navigation between English, as my primary academic language, and Georgian, as my fieldwork and native language. Furthermore, and more importantly, it meant continuously translating meanings, definitions and key organising themes from my project to my informants and back. As mentioned previously, I have tried to somehow bring these two types of 'translation' together by attempting to also publish in Georgian and by questioning and remaking Western concepts according to fieldwork experiences. However, I have also learnt that the transnational/Western academic context is very consistent in disciplining scholars to translate results back into established concepts and theories. Publishing in transnational, primarily English language journals certainly requires a thorough engagement with Western concepts and literature, leaving marginal space for conceptual innovations that would draw on Eastern empirical settings. However, this does not mean we should not push ourselves hard to find ways out and around the named challenges, and the following and concluding section attempts to do so.

Discussion: on how to steal time?

Research ethics is about how we navigate fieldwork, which principles are leading our interpersonal encounters, and how we respect contextual norms of social exchange and solve possible conflicts between diverse ethical pressures. Our selected personal experiences from Eastern Europe and Eurasia demonstrate that ethical challenges are hardly limited to questions of informed consent and the like but also require sensitivity towards global hierarchies and dependencies. Questions about ethical research positions are closely linked to questions of decolonisation and de-Westernisation. Knowledge production worldwide is enabled by predominantly Western European and North American resources and shaped by Western understandings of not just what ethics means but academic knowledge in general. Career prospects or sometimes simply day-to-day work and survival of academics in many parts of the world, particularly doctoral and early career researchers, are often dependent on Western academic standards and resources. Such dependency then poses limits to how far one can contribute personally to de-Westernisation and decolonisation of science structures and epistemological foundations while struggling to remain part of the academic system.

One of the clearest ethical challenges we have faced in our research, relating to Eurocentrism and coloniality of knowledge production, was the question of benefiting research participants and partners. The small contribution that our research could possibly make to knowledge production would hardly affect respective societies let alone the majority of research participants themselves. The feeling of not giving back remains a severe challenge. Limited opportunities for our research partners to contribute to the research process over the longer term are also a further manifestation of hierarchies in the production of knowledge. Research participants' inclusion in the research process usually ends with the data collection stage. They are mostly unable to participate in the analytical and reflexive process after the empirical phase and have limited or no access to research outcomes written in English or other languages, let alone visit scientific presentations. Developing solutions for transferring knowledge to where it is actually needed demands a lot of extra energy, critical reflection on standard routes, the power to change them and, most of all, time.

Time is an indisputable prerequisite, which under the conditions of capitalist economisation of science turns increasingly into a scarce resource. Time is necessary to get to know social and political contexts and their trans-scalar constitution comprehensively. It is also an indispensable feature for establishing the confidence and trust that enable us to conceptualise and realise research as a process of mutual (transnational) encounter. We see this as a proper way to initiate a decentred and shared process of knowledge production which, on the one hand, contributes to academia and, on the other hand, to research participant's experiences and life-worlds.

One lesson we draw is that we have to go beyond observing the subversion of neoliberal logics elsewhere and start participating actively in this process within academia, for example, by 'stealing' time: to spend it with research participants before and after data collection; to contribute to spaces of knowledge production which will not be counted in our formal performance; by attempting to make our research questions more relevant to local political contexts; by utilising our access to transnational academic space to bring up overlooked problems, concepts and challenges; and by sharing our privileged access to the resources of Western academia. While being well aware of our limits, our own embeddedness in and dependency on Western resources, we cannot stop thinking of tactics for subverting the fundamentally Eurocentric knowledge production system and, thereby, challenging existing knowledge production geographies. Far from being an individual endeavour, the search for such tactics requires support and solidarity between academic status groups within interdisciplinary and transnational networks.

References

Abu-Lughod, J. (1975) 'The legitimacy of comparisons in comparative urban studies. A theoretical position and an application to North African cities', *Urban Affairs Quarterly*, 11 (1): 13–35.

Askins, K. (2018) 'Feminist geographies and participatory action research: Co-producing narratives with people and place', *Gender, Place & Culture*, 25 (9): 1277–1294, https://doi.org/10.1080/0966369X.2018.1503159.

Askins, K. and Blazek, M. (2017) 'Feeling our way: academia. Emotions and a politics of care', *Social & Cultural Geography*, 18 (8): 1086–1105, https://doi.org/10.10 80/14649365.2016.1240224.

Bell, K. (2014) 'Resisting commensurability: Against informed consent as an anthropological virtue', *American Anthropologist*, 116 (3): 511–522.

Belur, J. (2014) 'Status, gender and geography: Power negotiations in police research', *Qualitative Research*, 14 (2): 184–200.

Berdahl, D. (2010) 'Introduction: An anthropology of postsocialism', in D. Berdahl, M. Bunzl and M. Lampland (eds) *Altering states. Ethnographies of transition in eastern Europe and the former Soviet Union.* Ann Arbor: University of Michigan Press, pp. 1–13.

Bohnsack, R., Pfaff, N. and Weller, W. (2010) *Qualitative analysis and documentary method in international educational research.* Opladen: Barbara Budrich Publishers.

Bondi, L. (2005) 'The place of emotions in research: From partitioning emotion and reason to the emotional dynamics of research relationships', in J. Davidson, L. Bondi and M. Smith (eds) *Emotional geographies.* Aldershot: Ashgate.

Brandtstädter, S. (2007) 'Transitional spaces: Postsocialism as a cultural process: Introduction', *Critique of Anthropology*, 27 (2): 131–145. https://doi.org/10.1177/03 08275X07076801.

Bruns, B. and Zichner, H. (2009) 'Übertragen – Übersetzen – Aushandeln? Wer oder was geht durch Übersetzung verloren, oder kann etwas gewonnen werden? '*Social Geography*, 4: 25–37, http://www.soc-geogr.net/4/25/2009/sg-4-25-2009.pdf (accessed 15 March 2010).

Buchowksi, M. (1996) 'The shifting meaning of civil and civic society in Poland', in E. Dunn and C. Hann (eds) *Civil society challenging Western models.* London: Routledge, pp.79–98.

Burawoy, M. and Verdery, K. (eds) (1999) *Uncertain transition: ethnographies of change in the postsocialist world.* Lanham, MD: Rowman & Littlefield.

Chakrabarty, D. (2000) *Provincializing Europe. Postcolonial thoughts and historical difference.* Princeton: Princeton University Press.

Cheterian, V. (ed) (2013) *From Perestroika to Rainbow revolutions.* London: Hurst & Company.

De Soto, H.G. (2000) 'Crossing Western boundaries: How East Berlin women observed women researchers from the West after Socialism 1991-1992', in H.G. De Soto and N. Dudwick (eds) *Fieldwork dilemmas. Anthropologists in postsocialist states.* Madison: The University of Wisconsin Press, pp.74–99.

Dudwick, N., Gomart, E., Marc, A. and Kuehnast, K. (2003) *When things fall apart: Qualitative studies of poverty in the former Soviet Union.* Washington, D.C.: The World Bank.

Escobar, A. (2007) 'Worlds and knowledges otherwise', *Cultural Studies*, 21: 179–210, https://doi.org/10.1080/09502380601162506.

Garfinkel, H. (1967) *Studies in ethnomethodology.* Englewood Cliffs, NJ: Prentice-Hall.

Gentile, M. (2018) 'Three metals and the "post-socialist city": Reclaiming the peripheries of urban knowledge', *International Journal of Urban and Regional Research*, 42 (6): 1140–1151. https://doi.org/10.1111/1468-2427.12552.

Glasius, M., de Lange, M., Bartman, J., Dalmasso, E., Del Sordi, A., Lv, A., Michaelsen, M. and Ruijgrok, K. (2018) *Research, ethics and risk in the authoritarian field.* Palgrave Macmillan, https://link.springer.com/content/pdf/10.1007%2F978-3-319-68966-1.pdf (accessed 06 June 2018).

Glück, A. (2015) 'De-Westernisation. Key concept paper', http://eprints.whiterose. ac.uk/117297/1/Glueck%202016_De-Westernisation.pdf (accessed 22 June 2018).

Hann, C. and Dunn, E. (1996) *Civil society: Challenging Western models*. London: Routledge.

Herzfeld, M. (2010) 'Foreword', in M. Bunzl (ed) *On the social of postsocialism. Memory, consumption, Germany. Daphne Berdahl*. Bloomington and Indianapolis: Indiana University Press, pp.vii–x.

Holston, J. (2008) *Insurgent citizenship. Disjunctions of democracy and modernity in Brazil*. Princeton: Princeton University Press.

Hörschelmann, K. (2002) 'History after the end: Post-socialist difference in a (post) modern world', *Transactions of the Institute of British Geographers*, 27 (1): 52–66.

Hörschelmann, K. and Stenning, A. (2008) 'Ethnographies of postsocialist change', *Progress in Human Geography*, 32 (3): 339–361, https://doi.org/10.1177/ 0309132508089094.

Humphrey, C. (2002) *The unmaking of Soviet life: Everyday economies after socialism*. Cornell: Cornell University Press.

Jones, S. (2013) *Georgia. A political history since independence*. London and New York: I.B. Tauris.

Kaufmann, W. (2009) 'Der weite Weg zur Zivilgesellschaft', *Kauskasus. Aus Politik und Zeitgeschichte*, 13: 12–18.

Knoblauch, H. (2008) 'Sinn und Subjektivität in der qualitativen Forschung', in H. Kalthoff, S. Hirschauer and G. Lindemann (eds) *Theoretische Empirie – Zur Relevanz qualitativer Forschung*. Frankfurt am Main: Suhrkamp, pp. 210–233.

Lassiter, L. E. (2005) *The Chicago guide to collaborative research*. London: University of Chicago Press.

Maldonado-Torres, N. (2004) 'The topology of being and the geopolitics of knowledge', *City*, 8: 29–56. https://doi.org/10.1080/1360481042000199787.

Mannheim, K. (1980/1924) 'Eine soziologische Theorie der Kultur und ihrer Erkennbarkeit (Konjunktives und kommunikatives Denken)', in D. Kettler, V. Meja and N. Stehr (eds). *Karl Mannheim. Strukturen des Denkens*. Frankfurt am Main: Suhrkamp, pp. 155–322.

Michailova, S. and Clark, E. (2004) 'Doing research in transforming contexts: themes and challenges', in E. Clark and S. Michailova (eds) *Fieldwork in transforming societies. Understanding methodology from experience*. Basingstoke: Palgrave Macmillan, pp.1–18.

Mignolo, W. (2002) 'Geopolitics of knowledge and the colonial difference', *South Atlantic Quarterly*, 101 (1): 57–96.

Mühlfried, F. (2006) *Postsowjetisches Feiern. Das Georgische Bankett im Wandel*. Stuttgart: ibidem.

Müller, M. (2020) 'In search of the global east: Thinking between north and south', *Geopolitics*, 25 (3): 734–755.

Müller, K. (2013) 'Yet another layer of peripheralization. Dealing with the consequences of the Schengen treaty at the edges of the EU territory', in A. Fischer-Tahir and M. Naumann (ed) *Peripheralization: the making of spatial dependencies and social injustice*. Heidelberg: Springer VS, pp.187–206.

Mullings, B. (1999) 'Insider or outsider, both or neither: Some dilemmas of interviewing in a cross-cultural setting', *Geoforum*, 30: 337–350.

Nodia, G. (2013) 'The record of the Rose Revolution. Mixed but still impressive', in V. Cheterian (ed) *From Perestroika to Rainbow Revolutions*. London: Hurst & Company, pp. 85–115.

Nohlen, D. (2002) *Kleines Lexikon der Politik*. München: Beck.

Pilz, M. (2019) *Das ist nicht georgisch! Postsozialistische Ausgrenzungspolitiken*. Berlin: Humboldt-Universität zu Berlin, https://edoc.hu-berlin.de/handle/18452/20882.

Radcliffe, S. (2017) 'Decolonising geographical knowledges', *Transactions of the Institute of British Geographers*, 42. 329–333.

Ries, N. (2000) 'Ethnography and post-socialism. Foreword', in H.G. De Soto and N. Dudwick (eds) *Fieldwork dilemmas. Anthropologists in postsocialist states*. Madison: The University of Wisconsin Press, pp. ix–xi.

Roberts, S. (2012) 'Research in challenging environments: The case of Russia's 'managed democracy'', *Qualitative Research*, 13 (3): 337–351. https://doi.org/10.1177/1468794112451039.

Robinson, J. (2011) 'Cities in a world of cities: The comparative gesture', *International Journal of Urban and Regional Research*, 35 (1): 1–23.

Roth, W.-M. (2005) 'Ethik als soziale Praxis: Einführung zur Debatte über qualitative Forschung und Ethik', *Forum: Qualitative Social Research*, 6 (1): Art. 9, http://www.qualitative-research.net/index.php/fqs/rt/printerFriendly/526/1138.

Schäuble, M., Rakowski, T. and Pessel, W.K. (2006) 'Doing fieldwork in Eastern Europe: Introduction', *Anthropology Matters*, 8 (1): 1–6, https://www.anthropology matters.com/index.php/ anth_matters/article/view/70/136 (accessed 31 May 2018).

Scheper-Hughes, N. (1995) 'The primacy of the ethical: Propositions for a militant anthropology', *Current Anthropology*, 36 (3): 409–420.

Soulsby, A. (2004) 'Who is observing whom? Fieldwork roles and ambiguities in organisational case study research', in E. Clark and S. Michailova (eds) *Fieldwork in transforming societies. Understanding methodology from experience*. Basingstoke: Palgrave Macmillan, pp.39–56.

Steger, T. (2004) 'Identities, roles and qualitative research in Central and Eastern Europe', in E. Clark and S. Michailova (eds) *Fieldwork in transforming societies. Understanding methodology from experience*. Basingstoke: Palgrave Macmillan, pp.19–38.

Stenning, A. and Hörschelmann, K. (2008) 'History, geography and difference in the post-socialist world: Or, do we still need post-socialism?', *Antipode*, 40 (2): 312–335, https://doi.org/10.1111/j.1467-8330.2008.00593.x.

Streiner, D.L. and Sidani, S. (eds) (2010) *When research goes off the rails: Why it happens and what you can do about it*. New York: Guilford.

Suyarkulova M. (2018) ''Renegade research': Hierarchies of knowledge production in Central Asia', *OpenDemocracy*, https://www.opendemocracy.net/od-russia/mohira-suyarkulova/renegade-research?fbclid=IwAR3wZX5Pyla-GJ4iStPUL6uc ZkhUdgIkml_2zVLz-t3tgRTscuvRTZFrDzk (accessed 20 December 2018).

Thelen, T. (2011) 'Shortage, fuzzy property and other dead ends in the anthropological analysis of (post)socialism', *Critique of Anthropology*, 31 (1): 43–61. https://doi.org/10.1177/0308275X10 393436.

Timár, J. (2004) 'More than 'Anglo-American', it is 'Western': Hegemony in geography from a Hungarian perspective', *Geoforum*, 35 (5): 533–538. https://doi.org/10.1016/j.geoforum.2004.01.010.

Tlostanova, M. (2015) 'Between the Russian/Soviet dependencies, neoliberal delusions, dewesternizing options, and decolonial drives', *Cultural Dynamics*, 27: 267–283. https://doi.org/10.1177/0921374015585230.

Tlostanova, M. (2018) 'The postcolonial and the postsocialist: A deferred coalition? Brothers forever?', *Postcolonial Interventions*, III (1): 1–37.

Trubina, E. (2018) 'Comparing at what scale? The challenge for comparative urbanism in Central Asia', in P. Horn, P.A. d'Alencon and A.C. Duarte Cardosa (eds) *Emerging urban spaces. A planetary perspective*. Cham: Springer International, pp. 109–127, https://link.springer.com/chapter/10.1007/978-3-319-57816-3_6 (accessed 17.04.2019).

Tuvikene, T. (2016) 'Strategies for comparative urbanism: Post-socialism as a deterritorialized concept', *International Journal of Urban and Regional Research*, 40 (1): 132–146. https://doi.org/10.1111/1468-2427.12333.

Verdery, K. (1996) *What was socialism, and what comes next?* Princeton: Princeton University Press.

Von Unger, H., Dilger, H. and Schönhuth, M. (2016) 'Ethikbegutachtung in der sozial- und kultur-wissenschaftlichen Forschung? Ein Debattenbeitrag aus soziologischer und ethnologischer Sicht', *Forum Qualitative Sozialforschung/Forum: Qualitative Social Research*, 17 (3): Art. 20, http://nbn-resolving.de/urn:nbn:de:0114-fqs1603203 (accessed 17 April 2019).

Wiest, K. (2012) 'Comparative debates in post-socialist urban studies', *Urban Geography*, 33 (6): 829–849, https://doi.org/10.2747/0272-3638.33.6.829.

7 Sensitive topics in human geography

Insights from research on cigarette smugglers and diamond dealers

Bettina Bruns and Sebastian Henn

Introduction

Whether a research topic is considered sensitive or not is an essential question that is answered subjectively in different ways. In many cases, aspects that are considered sensitive include "taboo topics, topics associated with shame or guilt, and topics that generally reside in the private spheres of our lives" (Noland 2012: 3). There seems to be a general understanding among researchers that sensitive topics are of a private and personal character (Cowles 1988: 163; De Laine 2000: 67; Robertson 2000: 531; Wellings et al. 2000: 256; Corbin and Morse 2003: 336; Rosenbaum and Langhinrichsen-Rohling 2006: 404; Noland 2012: 4) and often concern 'intimate' aspects of one's own life (Corbin and Morse 2003: 336). The creation of taboos or attitudes of not disclosing certain things are, of course, not necessarily associated with illegal and controversial aspects; many studies on sensitive topics, however, deal with illegal behaviour (Rosenbaum and Langhinrichsen-Rohling 2006: 404) or 'contentious areas' (Barnard 2005: 2) thereof. In fact, research on "trauma, abuse, death, illness, health problems, violence, crime" (Campbell 2002: 33) can all be regarded as sensitive due to the shamefulness and possibly strong emotional responses from the subjects in addition to fears, e.g. of prosecution. In this context, De Laine (2000) refers to what she calls 'back regions' (p. 67) in the sense of private spaces where only insiders participate in personal activities.

Researching sensitive topics must be considered a very complex endeavour as participants may feel distressed when disclosing information due to concerns about social disapproval or censure (Sieber and Stanley 1988: 49; Wellings et al. 2000: 256; Noland 2012: 3) but also because they may feel threatened (Lee and Renzetti 1990: 512), e.g. when the researcher intrudes upon private spheres, delves too deeply into personal experiences, impinges on the interests of powerful persons or the exercise of coercion and control, or deals with aspects usually set apart (e.g. sacred aspects) (Lee and Renzetti 1990: 512). Research may also be considered as threatening when the individuals believe that confidentiality or anonymity could be compromised – something that, in the worst case, could have tremendous social, financial, legal, or political consequences for them (see Corbin and Morse 2003: 336).

DOI: 10.4324/9780429507366-7

Against this background, interviews on certain topics often "arouse powerful emotions" (ibid.: 337). So far, it has been suggested that research on sensitive topics typically exposes those participating in the research to considerable risks. However, it should not be ignored that research on sensitive topics may also cause physical, emotional or psychological distress to the researcher (see Liebling 1999: 158; Elmir et al. 2011: 12; Jewkes 2014: 387). For example, researchers may not be able to detach themselves from the field of study (Elmir et al. 2011: 15), or face 'vicarious traumatisation' or burnout (Dunkley and Whelan 2006), or receive information that she or he might have to pass on to the authorities for legal reasons, thus putting her or him as well as the research subject in quite an uncomfortable situation.

While an increasing number of studies from various disciplines have intensively addressed sensitive research contexts in recent years, there has been a clear lack of systematic analysis of the challenges that characterise such studies in the field of human geography. This contribution therefore aims to close this research gap by dealing with specific challenges that characterise geographical research by its sensitive context. For this purpose, we differentiate between challenges that occur while choosing an appropriate methodological framework and those that occur when it comes to protecting the research subjects' and the researchers' rights while conducting fieldwork. To achieve this, the paper is structured as follows: We start with a short overview of sensitive research topics in various disciplines suggesting that this research field is still in an early stage of development in geography. In the subsequent section, we distinguish sensitive research topics from related concepts – specifically, vulnerability and closed contexts. After that, we focus on the ethical challenges associated with the research of sensitive contexts. Specifically, we describe these challenges in an abstract manner before illustrating them in more detail using two case studies – cigarette smugglers and diamond traders – as examples. Finally, by decontextualising our findings, we derive recommendations for researchers who are either interested in studying sensitive topics in the future or have already been confronted with similar challenges.

Sensitive research contexts as a multidisciplinary research field

While we follow Lee and Renzetti (1990: 512) in the belief that sensitive topics potentially pose a substantial threat "for those involved, the emergence of which renders problematic for the researcher/or the researched the collection, holding and/or dissemination of research data," we also acknowledge that there are no objective criteria to determine whether a topic should be considered sensitive or not. Rather, sensitivity may only be evaluated in concrete settings by those involved in the research process (see Dickson-Swift et al. 2008b: 133). Whether something will be regarded as sensitive or not is, indeed, "socially influenced, culturally determined, and can be highly subjective to each individual at any given point in time" (Noland 2012: 4). In the end, "it is probably possible for any topic, depending on the context, to be a sensitive one" (Lee and Renzetti 1990: 512). Taking this aspect into consideration, it does not come as a surprise that sensitive topics have been dealt with in various disciplines.

The scientific debate on sensitive topics originates in health and nursing studies (see Cowles 1988; Kavanaugh and Ayres 1998; Anderson and Hatton 2000; Dickson-Swift et al. 2008a, 2008b; Elmir et al. 2011) that deal, for example, with HIV/AIDS (Davis et al. 2004), mental health issues, bereavement, fertility, abortion, miscarriage and terminal illnesses such as cancer (Alty and Rodham 1998; Davis et al. 2004). Further, in related disciplines, such as psychology (Sieber and Stanley 1988; Decker et al. 2011), criminology (Rosenbaum and Langhinrichsen-Rohling 2006; Jewkes 2014) and communication studies (Noland 2012), sensitive topics have played an important role for some time now. More recently, conceptual advancements, refined methodologies as well as the prominent role of transdisciplinary research, have resulted in a growing importance of research on sensitive topics in the humanities (e.g. Lee 1993, 1995; Lee and Lee 2012). In political and social sciences, for example, the sensitivity of research is associated with difficult and even dangerous circumstances that confront both researchers and the research subjects involved with legal and/or political consequences from participating in the research. Further, participants may also fear the social consequences of their commitment to research when deviant or illegitimate topics are in focus. Other studies have dealt with interpersonal violence (e.g. domestic violence, child abuse, sexual assault) (McCosker et al. 2001), intrapersonal violence (e.g. suicidal behaviour) as well as risk behaviour (e.g. drug use and abuse, various types of sexual behaviour) (Rosenbaum and Langhinrichsen-Rohling 2006: 404). Further, prison research (Jewkes 2014), research on sex communication (Noland 2012), perinatal loss (Kavanaugh and Ayres 1998), with homeless persons (Anderson 1996) or survivors of murder victims (Cowles 1988) deserve a mention in this context.

Over recent years, an increasing number of human geographers have also become interested in studying sensitive topics understood here as research on taboo subjects, outlawed or patterns of behaviour that are generally unaccepted, as well as illegal or illegalised actions. Researchers have focused, for example, on lived experiences and policy engagement with alcohol (Lawhon et al. 2014), historical geographies of motherhood (Moore 2009), on farmers in rural areas in developing countries (Scott et al. 2006), as well as on teaching children's geographies of sexualities in light of student and institutional expectations and evaluations (Hall 2020). Despite the growing interest of human geography in studying sensitive research topics, we not only agree with Lawhon et al. (2014) that geographers have "not yet fully engaged with the international literature on the ethical and methodological challenges of researching such topics" (p. 15) but also note that the overall number on publications on sensitive topics with an explicit geographical background is still limited today. At the same time, however, we assume that this topic will gain greater importance in the future. The reasons for this are manifold and include advances in qualitative research, an increasing interest in reflexivity, the increasing importance of ethnographic methods in the attempt to explain spatially relevant actions embedded in concrete social contexts, as well as new opportunities to "work in remote areas or areas that were previously 'off limits'" (Scott et al. 2006: 31).

Delineating sensitive topics from similar concepts

Research on sensitive topics is often conducted with people labelled as 'vulnerable' (see, for example, Melrose 2002; Benoit et al. 2005). Examples of the manifold groups that have been categorised as 'vulnerable,' 'difficult-to-access,' 'hidden' populations (Liamputtong 2007: 4) include child sex workers, gays and lesbians, refugees, bouncers, gang members, members of indigenous populations and people affected by stigmatised diseases such as schizophrenia, bulimia and anorexia (ibid.), but also pregnant women, mentally disabled persons and those who are economically or educationally at a disadvantage (Stone 2003). While these groups may appear to be quite different at first glimpse, what they have in common is that they are, at least in certain situations, "likely to be susceptible to coercive or undue influence" (Stone 2003: 149). To learn about these or similar groups, researchers typically delve deeply into the private spheres of those participating in the research and, by so doing, in many cases will be concerned with what could be called socially deviant settings. As the research subjects are often exposed to specific threats, most research on vulnerable persons can be clearly considered as research on sensitive topics. However, this should not create the impression that only vulnerable persons are the subject of research on sensitive topics. In fact, extremely wealthy and/or powerful actors can also be difficult to access and typically live quite covertly. For them, research may also become threatening if their interests (e.g. anonymity) are put at risk.

We also suggest distinguishing between what has been called closed contexts and sensitive topics. The term 'closed context' usually applies to authoritarian or illiberal states and places or to "settings that are predominantly defined by the prevalence of (...) acts of closure (...)" (Koch 2013: 390). In such contexts, the research process may be fraught by a certain 'culture of fear' (ibid.: 394). While this, of course, will have an effect on the way in which a project is designed and the sets of methods to be applied, surprisingly questions of how to conduct field research in 'closed settings' largely remain unanswered in the literature (ibid.: 391). As research in non-democratic circumstances can imply significant risks to researchers and the ones involved in their research, doing fieldwork in 'closed contexts' can be regarded as a specific type of researching sensitive topics. This holds true, for example, when doing research in authoritarian states where the possible presence and interference of secret services raise significant concerns of practical and ethical nature (Gentile 2013: 426). While research in closed contexts thus clearly defines another subset of research with sensitive topics, of course, not every study on sensitive topics vice versa takes place in closed contexts in the above-defined sense.

Challenges related to researching sensitive topics

It is generally acknowledged that "ethics provide the basis for conduct in research" (Corbin and Morse 2003: 348). One of the main ethical research principles mentioned in the literature is that nobody should be harmed as the result of any research conducted (Sieber and Stanley 1988: 49; Melrose 2002: 343; Gentile 2013: 426f). This includes the anonymity of research participants and the strict confidentiality of data produced during the research process, such as recorded interviews. The secure handling of information about individuals collected during the research process is therefore of crucial importance when researching sensitive topics, as any disclosure could confront the participants with severe social, legal, financial, or political consequences (Corbin and Morse 2003: 336). These considerations also include the final publication of research results (Larossa et al. 1981: 311), which should be done in a way that avoids sensationalist media coverage (Melrose 2002: 343).

While the above aspects, of course, apply in general, they are even more important in the case of sensitive research environments because of the severe risks associated and the dangers for those involved as mentioned above (Sieber and Stanley 1988: 49). This is primarily the case as "privacy interests of the individual must be balanced against the interest of the society to be protected from crime and that for very serious crimes, breaching confidentiality may be appropriate" (Sieber and Stanley 1988). Not surprisingly, "socially sensitive information is the kind of research information to which law enforcement or other governmental agencies may most want access" (Sieber and Stanley 1988: 51).

From the previous considerations, it follows that research on sensitive topics is associated with two major challenges: first, researchers have to develop an adequate set of appropriate methodological instruments to be able to collect data under (sometimes extremely) difficult research conditions. Second, during data collection and in all subsequent steps, the individual rights of both the researchers involved and the research subjects must be upheld, which can be particularly difficult when the interests of the parties involved in the research conflict with each other. Against this background, both the following conceptual explanations and the subsequent field studies will focus on these two aspects.

Choosing an appropriate methodological framework

In order to be able to collect data in sensitive contexts, researchers in most cases need to develop a high level of trust with their research subjects (Bruns and Henn 2014: 8). Building contacts to potential gatekeepers is often very time consuming, which is the reason why projects typically include longer stays in the field, as illustrated by studies on Nicaraguan youth gangs (Rodgers 2007), on juvenile prostitution in England and Wales (Melrose 2002), on youth cultures against the background of political upheavals (Nilan 2002) as well as on private policing in South Africa (Diphoorn 2012) or on imprisoned young offenders (Hassan 2016).[1] The relationship established by researchers is by no means limited to trust alone, but rather in many cases also supported by the fact that participants have very specific reasons as to why they agree to an interview – hoping, for example, to gain certain advantages from it (e.g. making contacts or establishing quid-pro-quo situations). It is probably only in rare cases that close relations between researchers and research subjects lead to the fact that the researcher is no longer able to distance her or himself sufficiently from the object of investigation, which in the worst case, can make dubious practices more valuable or even acceptable. Rodgers (2007), for example, spent several months of ethnographic fieldwork with a violent Nicaraguan youth gang. During that time, he was not only a victim but also an offender of violence. He had to "participate in a range of violent and illegal activities, including gang wars, thefts, fights, beatings, fencing, and conflicts with the police, as a result of which" (Rodgers 2007: 12), and endured "a number of things (…), including being attacked, threatened, beaten up, knifed, shot at, and thrown out of a moving car" (ibid.). In retrospect, it appears that Rodgers (2007) justifies exposing himself to a very hostile environment and acting violently by the actual circumstances he had found himself in during his fieldwork. This is, of course, an extreme example of the extraordinary stress that a researcher may experience when dealing with sensitive topics. While this kind of research greatly reduces the academic void of reflecting the security-related risks of fieldwork, which up to now remains 'scarce, vague or tangential' (Gentile 2013: 426), it can also be considered as 'extremely unethical to conduct' (Anderson and Hatton 2000: 245) research if the risks incurred, exceed the benefits associated with it. Another interesting constellation concerns research aimed at uncovering illegal and/or morally reprehensible actions. Scheper-Hughes (2004), for example, partially worked with investigative journalists in her work on organ trafficking. She is concerned with a scientific (ethnographic) approach to criminal acts, for the research of which she has sometimes disguised herself as she would otherwise not have gained access as a scientist. Aware of the risks she takes, she does not, however, 'participate' in the same way as Rogers. In order to be able to choose an appropriate methodological approach that is ethically justifiable, the researcher obviously needs detailed information about her or his field of investigation that may, however, not be available to her or him in the run-up to fieldwork. Rather, in many cases, this information will only be developed

---

1 By contrast, studies with a background in psychology or health studies often rely on unstructured/narrative interviews (e.g. Cowles 1988; Kavanaugh and Ayres 1998; Anderson and Hatton 2000; Corbin and Morse 2003; Elmir et al. 2011).

during her or his own empirical research. In practice, this poses the danger that she or he will not be able to deal intensively with the challenges of the field and, in the worst case, expose her- or himself and others to unknown risks.

Protecting the research subjects' and researchers' rights

As we outlined above, research on sensitive topics is only feasible, successful and ethically justifiable if the research subject is protected against any risks arising from the intended or unintended disclosure of information. The researcher must therefore credibly convey to the research subject that she or he will not disclose the information obtained to third parties. During the research process, however, it may arise that, despite this assurance, the researcher feels obliged to disclose the data with explicit reference to the research subject. This may be due to legal provisions obligating the researcher to report certain information that has become known to the state authorities, as the researcher would otherwise be liable to prosecution (offense of omission). Special regulations that typically apply to clergymen or lawyers, which exempt them from the obligation to pass on information, typically do not apply to researchers. In Germany, for example, researchers become liable to prosecution if they learn of certain crimes but do not report them to the authorities. These crimes include treason, forgery of money and securities, murder and manslaughter, robbery or predatory extortion, causing an explosion or flood of explosives, forced prostitution and certain other forms of sexual exploitation. Particularly in authoritarian or non-democratic states ('closed contexts') researchers can find themselves in considerable difficulties if they are unfamiliar with the relevant regulations. All this implies that in order to protect themselves researchers should ideally have at least some basic knowledge about the criminal law at the place of their field research. This can be considered a major challenge as the legal systems of the nation-states differ considerably from one another and the understanding of the relevant regulations requires professional legal expertise.

Furthermore, research, while legally sound, can also raise ethical questions and/or also emotionally overwhelm the researcher, who may experience the information she or he has been given as socially and/or psychologically stressful and thus obliged to pass it on. If a researcher decides to disclose data after providing credible assurance of confidentiality, this might expose the researcher to consequences under civil law, as the research subject can claim a contractual penalty.

Approaching the challenges in practice – insights from two case studies

To illustrate the above-mentioned challenges in the field of human geography, we will deal with two case studies on cigarette smugglers and diamond traders in the following. In particular, we will discuss the peculiarities of these settings and any related challenges by referring to the question of choosing an appropriate methodological framework and protecting the research subjects' and researchers' rights during the fieldwork.

Case study I: Researching cross-border cigarette smugglers

Defining the sensitive character of the study

The first case study refers to small-scale cigarette smugglers from Poland who smuggle mostly cigarettes across the Polish-Russian border. The field-work took place in a small town of roughly 20,000 inhabitants close to a large border crossing point on the Polish side of the border. Since the collapse of socialism, this place has suffered strongly from the negative impacts associated with political transformation. High unemployment rates, low wage levels, a high share of poor people in the population and significant labour outmigration of the local population are the main factors characterising the weak economic situation of the town. All these factors have led to a great increase in the importance of the informal sector. Besides black labour, small-scale smuggling of cigarettes and alcohol from the Russian to the Polish side of the border has become an effective strategy for many inhabitants of the structurally weak and peripheral region to cope with existential fears. As the above-mentioned products differ significantly in price between Poland and Russia, for many of these inhabitants the border has turned into an important economic source. Border officials estimate that at least 85 percent of people crossing the border do so with the intention of smuggling goods into Poland (Bruns 2010: 88). For some of them, the income generated by smuggling is their only livelihood, while others combine smuggling with their official low-paid jobs to earn a living.

Smuggling goods is an illegal activity, although the smuggling of very small amounts of cigarettes is considered a minor breach of the law. Smugglers therefore have to make sure that they do not get caught with their goods by the border authorities; they must be careful when selling their goods in the street and also be alert that their neighbours do not become suspicious as far as their trading activities are concerned. Due to these factors, it is extremely important for them to minimise risks in order to generate sufficient income from their activities. They do this by building selective and strategic relationships with important actors within the smuggling ring such as customs officers, border guards, customs and co-smugglers. By being part of these essential social networks, they are able to gain access to necessary up-to-date local knowledge in order to be successful with their economic practices.

Although smuggling is morally legitimised to some extent by the local community and border authorities due to the obviously desperate economic situation of large parts of the population, it is still an illegal activity. Getting caught can have tremendous consequences for the smugglers as alternative sources of income are not easy to find. At the same time, the illegality makes smugglers feel ashamed of their economic activity. In fact, they are not proud of how they earn a living. Instead, they try to hide their activities and secretly exploit the border economically.

Under these circumstances, smugglers are very reluctant to provide outsiders with insights into their trading activities and to share sensitive knowledge with unknown people. Respondents often expressed their fears of losing their anonymity to the researcher when stating that they lived in a very small environment. They were afraid of the reactions within their social circles if their relations were released to the smuggling rings and the border authorities. An email sent by an informant whose brother was a smuggler and who had asked other smugglers whether they would disclose any information to the researcher illustrates this reluctance quite vividly: "Unfortunately, my brother could not manage to persuade anyone – none of his friends believe that an interview wouldn't be used to their disadvantage. They are terribly distrustful and don't want to get into trouble. Since a journalist showed up at the border, the customs officials tightened controls, it was a big scandal. Since then, no one wants to talk anymore. My brother also said that it is not willingly accepted in the smuggler's environment when someone talks about this issue with other people – these persons get into difficulties later" (information provided by email on 20 April 2006). Hence, smugglers are afraid of losing their anonymity, being excluded from social networks and facing financial damages when exposing their economic practices to third parties.

Choosing an appropriate methodological framework

The study followed three aims. First, the aim was to find out more about how smuggling is carried out and about the rules and rituals applied when crossing the border. Second, a particular focus was placed on the individual meanings assigned to smuggling by people facing poverty. And third, the implications of smuggling for regional development were investigated.

The starting point when designing the data collection was the fact that the research object, namely smuggling, is per se against the law. Due to the illegality of this activity, the participation in research on smuggling posed a substantial threat to the researched subject which turned the topic into a sensitive one. Therefore, the researcher conducted a year-long ethnographic fieldwork, carrying out participant observation and qualitative interviews with smugglers, but also with local authorities, border guards and customs officials. Data collection started with the most easily accessible interview partners from local administration such as the mayor, representatives from the employment office, the office of social work and public institutions like the library, the local newspaper and the Catholic Church. The aim of this first step was on the one hand to obtain general information about the local society and on the other hand to develop the researcher's social integration into the local community. Besides conducting expert interviews, analysing official data and statistics and undertaking participant observations at relevant locations such as marketplaces also took place during the first half of the fieldwork. Meanwhile, trusting relationships could be developed with numerous people and contacts established with smugglers through the snowball system. During the second fieldwork period, the project included problem-centred interviews with smugglers and participant observation during smuggling trips across the border to Russia and back. Right at the end of the field stay, expert interviews with representatives from the border guards and customs officials followed. This order proved to be useful. Had border authorities been interviewed first and had smugglers noticed the meetings with the researcher, they might have become suspicious and afraid that confidential information provided by them would be passed on to the customs office. All in all, the implementation of these methods was only possible because of the researcher's long presence in the research field.

Protecting the research subjects' and researchers' rights

As described above, smuggling is an economic practice, which because of its illegality and the shame of the people living from it, is carried out as secretly as possible. Smugglers risk a lot when sharing intimate information with strangers about their routines crossing the border: their anonymity, their social networks and, ultimately, their existential income. As the research location is quite small, people know each other. Hence, an ethical challenge for the researcher consisted in the extremely discreet and accurate handling of data and information provided by the respondents.

The fact that the researcher was a clear outsider helped to establish trust between her and the local population. A university background as justification for the presence of the researcher on site was accepted by the locals as plausible. The research was conducted in a transparent and honest manner, by making clear its content and aims and the researcher's professional and personal background. However, even more important for the development of an adequate field access than official explanations were the people's perception of the researcher's day-to-day conduct within the research field. The researcher was perceived as a harmless, friendly and very interested student who even had to be taught the commonly accepted rules and procedures at the border, as she did not have any idea about smuggling at the beginning.

The empirical work was successful because the researcher was able to convince the subjects that she was protecting the information provided to her. This was done by guaranteeing full anonymity and by promising not to pass on any given information to third parties during the whole research process. In addition, the researcher granted many smugglers' requests not to conduct the interviews at their homes but at neutral places for the sake of their anonymity. This ensured that people would not question the researcher's intentions at their neighbours' door. At the beginning of the project, the researcher reflected intensively about whether she would support and cover up illegal activities and whether this would go hand in hand with her 'ethical compass.' Apart from the fact that most smugglers that she dealt with carried only very small amounts of cigarettes, which did not exceed the quantity necessary for presenting a criminal act, she soon learnt that illegality and its legitimisation within the field in this case went hand in hand. Due to the small amounts of goods smuggled in the researcher's presence, she was never confronted with the ethical dilemma of noticing a crime and thus having to report it to the authorities, even though she had guaranteed her interview partners full anonymity.

Case study II: Conducting research on diamond traders

Defining the sensitive character of the study

Our second study focuses on spatial shifts in the trade of jewelry diamonds. More specifically, it aimed at analysing the economic changes at the three diamond trading hubs of New York, Antwerp and Mumbai. In order to better understand the extent to which research in this field represents research on a sensitive topic, and which ethical challenges researchers are confronted with when analysing actors in these sectors, some of the special features of the diamond sector will be characterised in more detail in the following (for more details on the social organisation of the diamond trade and the related social institutions, see Richman 2006; Henn 2012, 2019).

The starting point for further considerations is the fact that diamonds are extremely valuable goods which, due to their small size, can be transported over long distances without any problems (and quite unobtrusively). Especially in view of their high value, the diamond sector has traditionally been confronted with the question of how to avoid opportunistic behaviour on the part of individual players (e.g. exchange of valuable diamonds for less valuable ones) and how to ensure the reliability of expectations in the transaction process.

A set of common norms and values applying to all members in the sector as well as related, sanctioning mechanisms serve as an important governance structure. Further, in order to be able to engage in transactions more intensively, the actors must have a reputation from their reliability in past transactions. Typically, however, this reputation does not exist on an individual basis only, but rather extends to entire families (Shor 1989; Richman 2006). This ensures that any obligations that may arise from transactions are maintained irrespective of time, which can play an important role especially in those transactions involving goods of high value or the lending of credit. For new generations, this family-based reputation implies a reduction in transaction costs as they do not have to go through the difficult process of building trust in repeated transactions but can rely on the reputation of ancestors. At the same time, however, it presupposes that in order to be able to benefit from such reputation, future generations will continuously need to maintain such a good reputation through trust by keeping the rules in the transaction process.

Under these circumstances, individual actions that deviate from the gener-ally shared rules can cause sudden and permanent damage to the reputation of individuals and entire families. This can have considerable consequences for the defecting actor. In the case of particularly serious 'misdemeanors,' this can even result in her or his official exclusion from the trading network in question. Besides the family, the traders' business partners might also suffer the consequences with the degree of 'damage' being clearly influenced by how much they had known about such practices or to what extent they succeed in convincing others that they had not been any part of them.

It is clear from the above that diamond traders expose themselves and others to potentially significant social and economic risks when passing on information to the 'wrong' people. Consequently, the above points imply that diamond traders must be very careful about what they can tell and to whom. A particular risk for them is clearly posed by 'outsiders': In their case, not only are the motives for their actions often not sufficiently clear, but rather, there is also an increased uncertainty of expectations regarding the consequences of their actions, as they are not bound to sector-specific institutions and thus not subject to the generally accepted sanction mecha-nisms (the diamond sector has its own jurisdiction).

Similarly, when considering the fact that religious communities play a special role in the diamond sector, it becomes clear that relying on 'outsiders' poses additional risks as the absence of sanctions implies that there is no guarantee that religiously governed types of behaviour will be respected. For example, in Judaism, there is a highly important rule not to share any unnecessary information about third parties. If, however, third parties do not feel bound by this rule, this can result in information about the firm in question being uncontrollably leaked elsewhere – with unpredictable implications for the owner of the business. These aspects result in strong reservations about the diamond trading, which, from another perspective, might appear as a secluded social network. Without any doubt, this isola-tion can also be associated with the risk of becoming victim of a robbery or other crime. The lack of incentives for traders to operate outside the sector (e.g. marketing activities), the low average level of education and the main-tenance of traditional business practices while neglecting new practices (e.g. in the field of digitalisation) have further reinforced discretion in the sector. In addition, a significant proportion of traders engage in illegal business practices, such as the trade in uncertified diamonds, illegal employment and tax evasion (on the susceptibility of the sector to crime, see Siegel 2009: 159ff). The exposure of these practices by third parties could pose existen-tial problems for the actors.

Choosing an appropriate methodological framework

It was not the goal of the study to trace social networks or to uncover or even analyse illegal trading practices. Rather, it aimed at understanding how certain actors from India managed to develop a dominant position in the global diamond sector over only a few years and to force some established players out of the market. To this end, discussions with traders from different communities became essential. In particular, interviews with actors from a secluded community in India (the Palanpuri Jains) formed a background in order to learn about the factors that originally drove its members to move to Antwerp. It was also necessary to understand the ways in which the community was able to gain a foothold in Antwerp and the resistance its members encountered when entering the local market. In turn, interviews with the trader community (primarily Belgian Jews) were conducted in order to understand how local conditions had changed as a result of Indian traders entering the market.

Given the secretive nature of the diamond business, conducting the interviews was faced with the challenges of persuading interviewees to participate in the research, or making it clear to them that they were not putting themselves at any risk by participating, with the aim of obtaining information that was as useful as possible for the purpose of the research. To gain the trust of the interviewees, the researcher based his research for several months in the field. This allowed him to adapt to the conditions on site. It was, for example, essential that typical particularities of the sector – short-term arrangement of meetings in the high-security offices, arranging contacts without mentioning the name and/or firm of the next interviewee, dress code, etc. – were not called into question. Furthermore, repeated meetings with the same people helped to develop trust with a few people that later acted as import gatekeepers. Indeed, longstanding friendships did develop in this context; however, it was also possible to maintain a distance to the object of study because the researcher always made his interest in research clear as well as his university background, and emphasised the basic research character of the project to demonstrate that it was carried out on his own initiative and not on behalf of a third party (e.g. a competitor, bank or tax authority).

It also turned out that the research design that is chosen should not be seen as something static but should remain open to possible adjustments. In fact, as the exact research context was not entirely clear from the outset, some important decisions regarding the research design had to be made on site at short notice: In one case, the possibility arose to observe the interaction between the two communities in trading by participating in an observation in a diamond trader's office. This possibility had previously been regarded as unlikely and had therefore not been given any further consideration. When the opportunity arose by chance, it was immediately taken. Further, it should be mentioned that the original plan was to supplement the guided interviews with a quantitative survey in order to collect comparative data on the firms and their development. However, this plan soon had to be neglected for various reasons. First, the offices in which the diamond traders are mostly located are in high-security wings that would not have allowed questionnaires to be distributed. Second, it also became clear that the provision of information by the interviewees presupposes at least a minimum level of personal trust; the use of an anonymous questionnaire, whose author and objectives are difficult to assess a priori, would almost certainly not have been successful.

Protecting the research subjects' and researchers' rights

An analysis of illegal activities was not the subject of the project. Against this background, a deliberate attempt was made – wherever possible – to avoid addressing these issues in the interviews: In the guideline-based interviews, certain topics such as conflicts/blood diamonds, the certification of diamonds and/or import regulations/tariffs, illegal practices, etc. were deliberately not mentioned or at best only marginally mentioned, so as not to give the interviewees the impression that they were putting themselves at risk by answering. This did not pose any problems, as these aspects were not relevant for the study anyway. Furthermore, the names of other dealers from the sector and its immediate environment were deliberately not discussed. This was to demonstrate to the interview partners that they could also assume that their names would not be mentioned elsewhere. Illegal practices, such as illegal work, smuggling or tax evasion were mentioned in quite a few interviews. However, since the interviewees did not disclose any names, information on illegal activities always remained relatively abstract and could never be attributed to certain individuals. Furthermore, the most reported aspects clearly dated well back into the past and thus did not concern those individuals acting at the time of the interviews. As a result, the researcher himself was not confronted with the ethical dilemma that could potentially occur when conducting research on sensitive topics (see above).

The rights of the interviewees were safeguarded by the fact that the interviewees were assured that their names would not be mentioned in principle – neither to other interviewees nor in any publications.

While all efforts were made to emphasise that the project had no interest in analysing potentially illegal activities, various incidents demonstrated that some respondents nevertheless found themselves exposed to certain risks by participating in the interviews, highlighting the sensitive nature of the research environment:

- In one case, an interviewee followed the interviewer in the stairwell of the office complex after the interview had been completed, to ensure once again that the information collected would not be passed on. The fact that the information would remain exclusively with the interviewer had been explained to the interviewer repeatedly beforehand, at the end several minutes before leaving the office and also in response to a nervous question. Interestingly, the interview contained – at least from the interviewer's point of view – no information that could have been used even hypothetically against the interviewed partner.

- In various interviews, it became evident that some interviewees had deliberately and specifically gathered information about the researcher, and his intentions and questions before the interview took place. This was possible because contacts were usually mediated through the snowball system, and the contact with the intermediary gave the interviewed partners some kind of reinsurance.
- In addition, in several interviews, the question arose as to who had commissioned the research, and sometimes its answer was made a condition for the consent to an interview.
- In one case, one interviewee repeatedly put off the interviewer and repeatedly postponed the interview appointment. In the end, he did not attend an interview appointment.
- In one case, when the interviewer mentioned the origin of the diamonds, the interview was ended more or less abruptly.

Conclusion

Our study aimed at identifying and discussing the challenges that may arise when researching sensitive topics – both conceptually and, by referring to two case studies, from a practice perspective. To this end, we distinguished between the challenges that arise when choosing an appropriate methodological framework on the one hand and those challenges that arise when it comes to protecting the research subjects' and researchers' rights without threatening the latter in the case studies being researched because of specific circumstances. In particular, we have shown how sensitive topics can be researched in such a way to enable researchers in the respective contexts to reduce possible pitfalls as much as possible in order to gain access to reliable information in as many ways as possible.

With regard to the choice of an appropriate methodological framework, our study suggests that data collection in probably most cases presupposes that researchers succeed in developing trusting relationships with certain individuals who can at best act as gatekeepers and provide them with additional contacts and information. In fact, it is only possible in exceptional cases to start relevant field research from new. Building up trust is a time-consuming process that requires field visits over several months. Such stays generate opportunities for learning about the local conditions and for repeatedly meeting individuals, favourably also in informal settings that are conducive to the exchange of knowledge. When establishing contacts with communities, a process that may result in friendships, researchers must ensure that they maintain sufficient distance to the research object. In this context, making the institutional research context and/or the goal of the project as transparent as possible has proven to be beneficial; at the same

time, our case studies suggest that researchers should try their best to identify and accept the 'rules of the game' that characterise the object of analysis, for example, by complying with local dress codes and/or local practices.
As the research context will usually only partly be known from the outset,
the project in its conception will probably be based upon certain hypotheses
about how the field can be best accessed. It is possible that these turn out to
be unsound or inappropriate over the course of the project. In such cases,
researchers must be willing to revise some of their methodological tools or,
in exceptional cases, the entire research design at short notice. Sometimes,
this may imply the discontinuation of the entire project – a step that needs
to be considered as a risk factor when designing the project.

The protection of individuals' rights against the misuse or disclosure of
the data obtained in the field is probably the most important success factor for research in sensitive contexts. However, as we discussed in our conceptual part, in contrast to clergymen or lawyers, who in many countries
have to maintain the confidentiality of confessions by law, researchers are
legally obliged to report certain offenses to the authorities – regardless of
whether researchers have given their word to participants not to disclose
information. To avoid this dilemma, 'problematic' subjects can be deliberately excluded from research, as was the case in the second case study. The
researcher thus tacitly accepts a taboo, which may help to strengthen the
trust in his person. However, this is of course only possible in those cases
where the topic in question is not the central object of research. In any case,
an in-depth analysis of the legal system in the country where the fieldwork
will take place and the anticipation of possible difficulties in advance, as
well as an exchange of information with other researchers who have gained
experience in similar settings, may thus be a useful preparation for the fieldwork. Furthermore, inconvenient information may also come to light during the research process which does not have to be criminally proven and
therefore does not have to be disclosed. In this case, the researcher is faced
with a decision of conscience, which, even though it may cause practical and
emotional problems (Melrose 2002: 348), cannot be taken away from him.
Religious rules (e.g. in Judaism, there is a very broad debate relating to the
conditions under which information told in confidence must be divulged) or
ethical balancing rules can provide important orientation here.

While the nature of sensitive topics suggests that researching them will
probably always remain challenging, in the end much depends on the individual conduct of the researcher and the societal relevance of a certain topic
as to whether 'research on sensitive topics' is carried out in an ethical manner. In fact, we conclude with the words of Iphofen (2011: 176), who said that
"no topics are inherently 'unethical' in terms of our potential for scientific
interest in them, nor can any method be entirely proscribed. The ethical
choices to be made when adopting a method and/or a topic for study will
depend upon the scientific and social importance of the research and our
views on how the subjects to be studied should be treated."

Acknowledgements

We wish to thank Rav Zsolt Balla, Christine Henn, Nikolaus Knoepffler and Judith Miggelbrink for valuable suggestions that helped to sharpen our arguments.

References

Alty, A. and Rodham, K. (1998): The ouch! factor: Problems in conducting sensitive research. *Qualitative Health Research*, 8, 2, 275–282.

Anderson, D. G. (1996): Homeless women's perceptions about their families of origin. *Western Journal of Nursing Research*, 18, 1, 29–42.

Anderson, D. G. and Hatton, D. C. (2000): Accessing vulnerable populations for research. *Western Journal of Nursing Research*, 22, 2, 244–251.

Barnard, M. (2005): Discomforting research: Colliding moralities and looking for 'truth' in a study of parental drug problems. *Sociology of Health & Illness*, 27, 1, 1–19.

Benoit, C., Jansson, M., Millar, A. and Phillips, R. (2005): Community-academic research on hard-to-reach populations: Benefits and challenges. *Qualitative Health Research*, 15, 2, 263–82.

Bruns, B. (2010): *Grenze als Ressource. Die soziale Organisation von Schmuggel am Rande der Europäischen Union*. VS Verlag für Sozialwissenschaften.

Bruns, B. and Henn, S. (2014): Problem-centered interviews in sensitive contexts: Researching cigarette smugglers and diamond traders. *SAGE Research Methods Cases*, http://dx.doi.org/10.4135/978144627305013512941, accessed 11/23/2016.

Campbell, R. (2002): *Emotionally Involved: The Impact of Researching Rape*. New York: Routledge.

Corbin, J. and Morse, J. M. (2003): The unstructured interactive interview: issues of reciprocity and risks when dealing with sensitive topics. *Qualitative Inquiry*, 9, 3, 335–354.

Cowles, K. (1988): Issues in qualitative research on sensitive topics. *Western Journal of Nursing Research*, 10, 2, 163–170.

Davis, M., Bolding, G. and Hart, G. et al. (2004): Reflecting on the experience of interviewing online: perspectives from the Internet and HIV study in London. *AIDS Care*, 16, 8, 944–952.

Decker; N; Carter-Visscher, R., Bell, K. and Seifert, A. (2011): Ethical issues in research on sensitive topics: Participants' experiences on distress and benefit. *Journal of Empirical Research on Human Research Ethics*, 6, 3, 55–64.

De Laine, M. (2000): *Fieldwork, Participation and Practice: Ethics and Dilemmas in Qualitative Research*. London: Sage.

Dickson-Swift, V., James, E. L., Kippen, S. and Liamputtong, P. (2008a): *Undertaking Sensitive Research in the Health and Social Sciences: Managing Boundaries, emotions and Risks*. Cambridge: Cambridge University Press.

Dickson-Swift, V., James, E. L., Kippen, S. and Liamputtong, P. (2008b): Risk to researchers in qualitative research on sensitive topics: Issues and strategies. *Qualitative Health Research*, 18, 1, 133–144.

Diphoorn, T. (2012): The emotionality of participation in ethnographic fieldwork on private policing in Durban, South Africa. *Journal of Contemporary Ethnography*, 42, 2, 201–225.

Dunkley, J. A., Whelan, T. (2006): Vicarious traumatisation: current status and future directions. *British Journal of Guidance and Counselling*, 34, 1, 107–116.

Elmir, R., Schmied, V., Jackson, D. and Wilkes, L. (2011): Interviewing people about potentially sensitive topics. *Nurse Researcher*, 19, 1, 12–16.

Gentile, M. (2013): Meeting the 'organs': The tacit dilemma of field research in authoritarian states. *Area*, 45, 4, 426–432.

Hall, J. J. (2020): Approaching "sensitive" topics: criticality and permissibility in research-led teaching about children, sexualities, and schooling. *Journal of Geography in Higher Education*, 44, 2, 248–264.

Hassan, N. (2016): Surviving research on sensitive topics with young offenders. *Scottish Journal of Residential Child Care*, 15, 1, 102–114.

Henn, S. (2012): Transnational entrepreneurs, global pipelines and shifting production patterns. The example of the Palanpuris in the diamond sector. *Geoforum*, 43, 3, 497–506.

Henn, S. (2019): Diamanten in Antwerpen – transnationale Mobilität von Unternehmen, Wissen und Produktion (pp. 85–96). In: Gamerith, W. and Scharfenort, N. (eds): *Menschen, Migration und Mobilität*. Passau: Passauer Kontaktstudium Geographie.

Iphofen, R. (2011): *Ethical Decision Making in Social Research: A Practical Guide*. Palgrave Macmillan, Basingstoke.

Jewkes, Y. (2014): An introduction to doing prison research differently. *Qualitative Inquiry*, 20, 4, 387–391.

Kavanaugh, K. and Ayres, L. (1998): "Not as bad as it could have been": Assessing and mitigating harm during research on sensitive topics. *Research in Nursing and Health*, 21, 91–97.

Koch, N. (2013): Introduction – Field methods in 'closed contexts': undertaking research in authoritarian states and places. *Area*, 45, 4, 390–395.

Larossa, R., Bennett, L. A. and Gelles, R. J. (1981): Ethical dilemmas in qualitative family research. *Journal of Marriage and the Family*, 43, 2, 303–13.

Lawhon, M., Herrick, C. and Daya, S. (2014): Researching sensitive topics in African cities: reflections on alcohol research in Cape Town. *South African Geographical Journal*, 96, 1, 15–30.

Lee, R. M. (1993): Doing Research on Sensitive Topics. London, Newbury Park, New Delhi: Sage.

Lee, R. M. (1995): *Dangerous Fieldwork*. London: Sage.

Lee, Y.-O. and Lee, R. M. (2012): Methodological research on "sensitive" topics: A decade review. *Bulletin of Sociological Methodology/Bulletin de Méthodologie Sociologique*, 114, 1, 35–49.

Lee, R.M. and Renzetti, C. M. (1990): The problems of researching sensitive topics. *American Behavioral Scientist*, 33, 5, 510–528.

Liamputtong, P. (2007): *Researching the Vulnerable: A Guide to Sensitive Research Methods*. Sage: London.

Liebling, A. (1999): Doing research in prison: Breaking the silence?. *Theoretical Criminology*, 3, 2, 147–173.

McCosker, H., Barnard, A. and Gerber, R. (2001): Undertaking sensitive research: Issues and strategies for meeting the safety needs of all participants. *FQS – Forum: Qualitative Research*, 2, 1, Art. 22.

Melrose, M. (2002): Labour pains: Some considerations on the difficulties of researching juvenile prostitution. *International Journal of Social Research Methodology*, 5, 4, 333–351.

Moore, F. P. L. (2009): Tales from the archive: Methodological and ethical issues in historical geography research. *Area*, 42, 3, 262–270.

Nilan, P. (2002): 'Dangerous fieldwork' re-examined: The question of researcher subject position. *Qualitative Research*, 2, 3, 363–386.

Noland, M. (2012): Institutional barriers to research on sensitive topics: Case of sex communication research among university students. *Journal of Research Practice*, 8, 1, Article M2.

Richman, B. D. (2006): How community institutions create economic advantage: Jewish diamond traders in New York, *Law & Social Inquiry*, 31, 2, 383–420.

Robertson, J. (2000): Ethical issues and researching sensitive topics: Mature women and: 'Bulimia'. *Feminism & Psychology*, 10, 4, 531–537.

Rodgers, D. (2007): Joining the gang and becoming a broader: The violence of ethnography in contemporary Nicaragua. *Bulletin of Latin American Research*, 26, 4, 444–461.

Rosenbaum, A. and Langhinrichsen-Rohling, J. (2006): Meta-research on violence and victims: The impact of data collection methods on findings and participants. *Violence and Victims*, 21, 4, 404–409.

Scheper-Hughes, N. (2004): Parts unknown: Undercover ethnography of the organs-trafficking underworld. *Ethnography*, 5, 1, 29–73.

Scott, S., Miller, F. and Lloyd, K. (2006): Doing fieldwork in development geography: Research culture and research spaces in Vietnam. *Geographical Research*, 44, 1, 28–40.

Shor, R. (1989): Family: Diamond's critical connection. A look inside the tradition-bound world of diamond trading through the eyes of six families. *Jewelers Circular Keystone*, 190–206.

Sieber, J. E. and Stanley, B. (1988): Ethical and professional dimensions of socially sensitive research. *American Psychologist*, 43, 1, 49–55.

Siegel, D. (2009): *The Mazzel Ritual: Culture, Customs and Crime in the Diamond Trade*. Dordrecht: Springer.

Stone, T. H. (2003): The invisible vulnerable: The economically and educationally disadvantaged subjects of clinical research. *The Journal of Law, Medicine and Ethics*, 31, 1, 149–153.

Wellings, K., Branigan,P. and Mitchell, K. (2000): Discomfort, discord and discontinuity as data: Using focus groups to research sensitive topics. *Culture, Health & Sexuality*, 2, 3, 255–267.

8 Volunteer-practitioner research, relationships and friendship-liness

Re-enacting geographies of care

Matej Blazek and Kye Askins

Introduction

Human geography research unfolds from a variety of relationships. Certainly, since the 'cultural turn', and especially for those working from poststructural, feminist and more-than-representational perspectives, researchers are unlikely to approach participants solely as objects of inquiry. Instead, situated circumstances, embodied identities and complex moral and political positions shape research relationships, and impact ethical decisions (e.g. Malam 2004; Cloutier et al. 2015). Further, research participants are not the only ones with whom geographers engage in the field, as conducting research involves working with funders and research commissioners, gatekeepers, informants, intermediaries, academic and non-academic collaborators, members of audience and the wider public as 'recipients' of knowledge. Such an expansion of relationships and the intricacy of research practices they embed accordingly requires reframing the ethics of doing geographical research (Fuller and Askins 2010).

This chapter addresses ethical dimensions of a particular framing of research relationships, as we reflect on experience of doing geographical research in the roles of volunteer-practitioners. We understand this position as one where, central to fieldwork praxis, we adopt the role of practitioner in a helping or caring profession, working in a service-providing community-based organisation with people (some of) who(m) become involved in research projects as participants and in other ways. The priority of roles is important for defining this; our work as volunteer-practitioners takes primacy over research ambitions and we are fully committed to professional responsibilities. For participants, colleagues and others we meet during our work, we are volunteer-practitioners first and foremost, while we build on the experience, ties and activities in this role to assemble action-oriented research. This unfolds from the situated knowledge and crafted relationships with practitioner colleagues, those who use the provided service, and their communities. While research remains important to us as academics, its goals, practices or outcomes must not jeopardise the volunteer-practitioner work.

DOI: 10.4324/9780429507366-8

Volunteer-practitioner research is not a homogeneous position. The discussion in this chapter has been influenced by accounts from others doing research in diverse volunteer and/or practitioner roles (e.g. Cheung Judge 2016; Dickens 2017) and also by our experiences of working with practitioners, volunteers and those who engage with their services. These experiences range from conducting commissioned research as outsiders to the participating organisations (Smith et al. 2016), working as researchers but non-practitioners in service-providing organisations (Blazek 2014; Askins 2016), to volunteering work with no research element, including unwritten accounts of projects where our plans to set up research from volunteering praxis did not materialise. These histories are inherently woven into our academic identities and ethical philosophies; while in this chapter we primarily draw on two particular projects of collaboration, in which our positionalities have been clearly defined as volunteer-practitioners doing research. We offer some detail here at the outset, to contextualise our narratives and arguments.

For Kye, this is an ongoing participatory action research (PAR) project with a weekly drop-in service, facilitated by a local third sector organisation involved in community development in Glasgow, Scotland. The service is volunteer-led, with staff presence and support, with the number of volunteers varying as a formal commitment to weekly volunteering is not required. Volunteers and attendees alike stem from a diverse range of more-and-less marginalised groups, including parents (especially mothers), asylum seekers and refugees, people with mental and/or physical health issues, with insecure housing provision, long-term un- and under-employment, and drug and alcohol recovery issues. Kye began volunteering in May 2015, to explore the possibilities of developing a research focus and questions together with those involved with the drop-in, fully transparent with staff and volunteers as to her role as a volunteer intended to develop research. Volunteering has since expanded to other aspects of the service alongside the drop-in, as the research focus has evolved and participant-researchers developed its reach.

Matej's reflections draw on his doctoral fieldwork, which he completed between 2008 and 2011 as youth worker in a peripheral neighbourhood of Bratislava, the Slovak capital. He worked with a community-based charity organisation that provided a range of services for children and young people from the neighbourhood, ranging from a drop-in youth club, detached youth work, community development and advocacy activities, to participation in international youth exchanges. His role varied and evolved during his stay with the organisation, and the collaboration exceeded both the duration and activities of his PhD project. Central to his research focus was his role as a detached youth worker, where he spent three afternoons/week on average on the streets of the neighbourhood, engaging with children (as young as five years old) and young people (up to early 20s) in a non-directive way. From this positioning, he developed his ethnographic and participatory research inquiry carried out through a variety of activities (Blazek 2015).

We draw on these experiences to address a range of ethical propositions and challenges emerging from working in such roles. We reflect on how these roles envelop diverse relationships, framing both questions and inspirations regarding being with others. Subsequently, we also seek to suggest ways in which these kinds of engagement with others can be translated into our wider academic praxis and professional identities, beyond the specific context of volunteer-practitioner research fieldwork. As such, the chapter begins with a conceptual background for our discussion, outlining how fieldwork experiences relate to broader debates about research and professional ethics. The following sections map and explore a series of propositions unfolding from working at such an interface and challenges this role invokes, specifically around 'friendship-liness' and becoming friends. The concluding part argues for the promise of volunteer-practitioner research precisely because of and through its relationship focus, as a form of geographical praxis: as an ethical approach to developing academic identities through care, and being attentive to relationships in situated spaces of social encounters.

Geographical research and the helping and caring professions

A growing focus on inequality, interest in socio-spatial justice and commitment to work ethically with those experiencing marginalisation and exclusion (von Benzon and van Blerk 2017) bring geographers into closer contact with people who already have access, relationships and established modes of ethically informed praxis with such groups and individuals. The helping and caring professions[1] is an umbrella term for a variety of fields that involve direct engagement with people and address a particular aspect of their well-being. Broadly understood, the helping and caring professions include areas, such as social work, community work, youth work, counselling, nursing, care, coaching, some areas of education and ministry. Practitioners in such fields often help facilitate the first contact with research participants, and are also increasingly likely to provide inputs into ways in which research can and should be done, particularly regarding situated ethical decisions pertaining to a specific locality, community, individual or theme. Indeed, practitioners' knowledge of people's circumstances, expectations and concerns is irreplaceable, as routine academic protocols of research ethics cannot foresee these important factors in their entirety.

Practitioners and volunteers in service-providing organisations are also gaining greater roles in increasing academic agendas around research

---

1 We recognise that this is a contested term, especially with regard to the power relations caught up in 'helping' and 'caring for'. However, this umbrella term is widely used among such professionals, and we echo their tendency here.

impact and social relevance. Along with helping establish *where*, *with whom* and *how* research can take place, their experience and insights can be pivotal for determining what outcomes might be generated that would be of relevance to participants and their communities (Fuller and Askins 2010). In a critical framing of research impact that is shaped as a *two-way* process of engagement between researchers and non-academic partners, as emerging *throughout* (rather than after) the research process and at a variety of scales (Pain et al. 2011), the position of frontline practitioners and their close collaboration with researchers is crucial. Indigenous and anti- and post-colonial scholars have long pointed out the explicit and hidden uneven power relations that shape the initial encounter of researchers and research participants, and ongoing relations in and beyond the field (Denzin et al. 2008; Tuhiwai Smith 2012; TallBear 2014).

That is, the nature and extent of collaboration are as dynamic as the circumstances of the fieldwork, and the boundaries between working *for*, working *through*, working *alongside* and working *as* a volunteer-practitioner become blurred. The need to take on additional roles and adopt a fluid approach is characteristic of research evolving through close engagement with others. As such, volunteer-practitioner research positions bear similarities with those of ethnography (Conradson, 2003; Klocker, 2015), activism-research (Chatterton et al. 2007; Taylor 2014) or research with friends (Ellis 2007; Hall 2009), and the differences between the connections across all these positions can be rather elastic. Yet, while many ethnographers take on additional roles in order to create a pattern of meaningful presence and encounters in the field, the researcher agenda remains prioritised and vindicates the presence in the field. As volunteer-practitioners, the provision of professional service marks the purpose of presence in encounters with research participants, taking primacy in making everyday decisions. Research may be put at risk for the sake of maintaining professional conduct as a volunteer-practitioner, and volunteer-practitioner work must not be limited by the research agenda. Like activist research, the volunteer-practitioner approach pursues an agenda of social change but embodied relationships with others are negotiated, supervised and shaped through the professional framework of organisations through which research/professional praxis takes place. This framework includes goals and objectives, and more detailed aspects of the code of conduct, including pre-given ethical principles, forms of accountability and procedures of supervision. Thus, unlike in activism research, additional formal or informal training and qualifications might be required. And, in common with research working with friends, volunteer-practitioner research is centred round close, caring and affirmative relationships between researcher and participant. However, there are different boundaries regarding whether/how these relationships may/not span beyond field encounters, governed by the professional conduct stipulated by an organisation. Limits might be posed as to what is and is not ethical/permissible in relationships, so the service provision can be sustained.

Geography and relational ethics

Discussions in geography about research ethics often revolve around the issue of institutional versus relational ethics. Institutional (also procedural or formal) ethics is a dimension mandated by academic review boards, to govern and monitor how research procedures address matters such as informed consent, confidentiality, non-intrusiveness, respect and dignity, non-deception or work with 'vulnerable' participants. Guidelines are envisaged to prevent harm and exploitation of research participants, predominantly within a narrow understanding of academic researchers 'doing' research on individuals/publics (as objects of study). This framing has been widely problematised, particularly for its insistence of replicability and predictability of research actions and relationships, and the focus on protection of researchers and academic institutions rather than on the well-being of participants (Cahill 2007). Ellis (2007) also highlights that academic ethical codes operate on the premise that researchers and their participants have no relationships outside the scope of data generation, replicating the medical model of science with human participants (Dyer and Demeritt 2009), whereas the nature of human geography fieldwork places embodied relationalities at the heart of research encounter (Cloutier et al. 2015).

Relational ethics is a wider concept entailing a number of often disparate perspectives. An important source is the care ethics feminist philosophy (Gilligan, 1982) that articulates the self as a set of asymmetric connections and inter-dependencies with others. Ethical decisions cannot be generated from the realm of researcher autonomy, as the researcher's subjectivities evolve from and through relationships with others. It is thus impossible to design a definitive script of all fieldwork procedures. Moreover, indigenous and anti- and post-colonial scholars urge that ethical protocols must be foregrounded *as socio-culturally constructed*, and also seen as iterative processes rather than one-off provisions because circumstances of research encounters and relationships inevitably change (e.g. Kovach 2010).

However, even if relational ethics is always situated, it is more than situational (Ellis 2007). The concept rarely refers to exclusively ad hoc decisions responding to individual events. Mason's (2015) discussion of participatory ethics, as one approach to relational ethics, argues that it is "an ethical stand against neutrality which aims to generate social change: it is unethical not to take action to challenge oppressions which are encountered during the research" (500). In other words, relational ethics features researchers' (and participants') political standpoints and integrate them with situated ethical actions of fieldwork and institutional contexts (Cahill 2007). As modalities of relational ethics, Cahill (2007) thus writes about 'ethics of inclusion' as a vehicle of social transformation and Askins (2007) theorises 'ethics of encounter' as an effort of resistance to resolution and closure. Similarly, Dickens and Butcher (2017) articulate 'ethics of recognition' in which the priority of anonymity and 'protection' of participants is superseded by their choice to become visible as political actors, while Richardson (2015) develops a comparable argument emphasising the importance of 'safe spaces' in which research stories are told.

Another aspect of relational ethics is its temporality. Where academic institutional reviews are concerned by the ways of recruiting research participants, processes of data generation and the confidentiality of research outputs, relational ethics emphasise how research practices are grounded in long-term, ongoing and often compound and evolving relationships with and beyond research participants, and of the long-term impact generated by research (Evans 2016). Reflecting on this, Zhang (2017: 147) proposes "reciprocating kindness in humble but practical ways" as an underlying strategy of ethical research approach, in contrast with 'blind compliance' with 'prescriptive ethics' (ibid.). Likewise, Bignante et al. (2016) advocate long-term, tacit and reciprocal relationships in communities as a cornerstone of research ethics, as such relationships have more capacity to accommodate particular dilemmas of situated research practices. Rather than being a one-off tick-box exercise producing a template framework for field conduct, relational ethics requires iterative praxis, ongoing reflexivity and investment in a two-way engagement with others.

Ethics in the helping and caring professions

As mentioned above, the helping and caring professions are constituted by a variety of fields, centred around direct engagement with people and with specific well-being goals. The term is loosely defined, but what separates it from other forms of interpersonal professional work such as medicine or (formal) education is the importance, intensity and authenticity of relationships between practitioners and service users (Skovholt and Trotter-Mathison 2016). Congruent relationships are vehicles of an efficacious service provision and their development is therefore at the core of professionals' training, virtues and focus.

As such, reflexive notions of ethics constitute cornerstones of field praxis (Corey and Corey 2014). Although most of the helping and caring professions have explicit or even statutory ethical codes of conduct, emphasis on the serendipity of relationships with service users and the reflexive nature of the contact work prevents practitioners from relying exclusively on prescribed sets of ethical principles. Rather, generic principles are expressly situated through reflexivity and care in evolving relationships with people from different backgrounds and in diverse circumstances. Relationships themselves are the means through which outcomes of professional work are achieved. Further, the ethical responsibility of helping and caring professionals stretches beyond service users, to their families and communities, to other practitioners, and to the wider public, highlighting the complex inter-dependent positioning of professionals and their wider responsibilities in multiple social contexts (Corey et al. 2014).

In our experience, the helping and caring professions operate with ethical values and principles that are in many cases similar to those of social research. As in research, avoidance of harm (non-maleficence) is the principal issue, although some professions, such as psychotherapy, rearticulate this as the prohibition of client's exploitation (Bond 2015). Confidentiality, non-deception, non-coercion, respect and dignity and the voluntary nature of participation are among central principles applied consistently across the spectrum of the helping and caring disciplines. However, the latter differ from established academic research praxis in one key respect, in that ethics are systemically positioned in institutional and professional infrastructures. One core aspect of this, responding to the ever-changing circumstances of interpersonal relationships, is the use of clinical supervision (cf. Proudfoot 2015). Whereas academic supervision is about the relationship between the researcher and their research (and data, particularly), clinical supervision is focused on the relationship between the practitioner and their participants, particularly the impact of their relationship on each (Cornforth and Claiborne 2008). Largely a compulsory practice in professions and disciplines where the well-being of practitioners and participants is of concern, it is conducted by trained experienced practitioners who are normally not otherwise involved in the activities of the supervisee. Consideration of ethics in the field is thus not vested with an isolated practitioner or detached institutional board, but with a fully engaged, informed and experienced professional who reflects on the practitioner-participant relationship on a regular basis.

Doing research as volunteer-practitioners

Our work as volunteer-practitioner researchers is driven by a number of factors. Commitment to social and spatial justice was and remains a motivation to work in non-exploitative ways with those who are marginalised in society and academic (field)work, whether it is people from the South of Glasgow experiencing difficult situations or children and young people from a deprived neighbourhood of Bratislava. Our professional circumstances during respective projects provided us with time to do this kind of research, requiring long-term regular commitment and involvement in non-research activities such as training or staff meetings: Matej had time to complete a lengthy ethnographic study as part of his PhD; Kye moved into a post with emphasis on research activities. We each found our collaboration with the hosting organisations and the engagement with other practitioners, volunteers and service users to be emotionally fulfilling while challenging; we experienced compassion, sense of purpose and professional as well as personal growth while working with them (Askins and Blazek 2017). But as a crucial underlying factor, doing research as volunteer-practitioners resonated for us with a politics of care that we both see as central to who we strive to be as academics – to our professional and personal identities beyond the narrow scope of research fieldwork:

> [W]e understand care as meeting the needs, and maintaining the worlds, of ourselves and others (Sevenhuijsen, 1998). We see care, then, as highly relevant to contemplating our places in academia. Our move towards a *politics* of care is thus about more than maintaining; rather we seek to *work towards* fair and sustainable relations in academia as related to wider societies.
>
> (Askins and Blazek 2017: 1089)

Being volunteer-practitioner researchers is thus significant at three levels, each with concomitant ethical considerations. First, it enables us to do in-depth research on themes of socio-spatial in/justice. Second, it enables us to develop research praxis that is underpinned by relationships of care, including with people outside our close circles. Third, it enables us to conduct a type of research where the process itself is intended as a tool of social and political transformation.

In the rest of the chapter, we seek to unpick these rationales as they intersect with a variety of tensions and dilemmas. We pay attention to how the character of researcher and volunteer-practitioner roles match and clash, and we discuss how critically reflecting on the role of *underlying and evolving relationships* can help us to navigate ethical praxis. Finally, we suggest how these reflections can help us rethink what academic work could be, across a wider range of contexts.

Non-maleficence and beneficence

As researchers, we are trained not to cause harm (non-maleficence). Both institutional and relational frameworks of ethics are principally concerned with the consequences of our actions on research participants and prevent us from acting in a way that would be detrimental for them. A range of research philosophies, such as PAR, feminist, post-colonial or activist research, seek to go further and formulate goals or principles grounded in affirmative social transformation. Still, 'doing good' (beneficence) in relation to participants often largely remains a secondary goal next to doing research; albeit with a range of aims, activities and fluid approaches to 'give back' and benefit participants, researchers engage with participants (arguably) primarily in order to do research (Taylor 2014).

Our point here is that as volunteer-practitioners, engagement with participants is first and foremost about their benefits and gains, through an ethics of being responsive to the needs and challenges faced by service users. The rationale for all activities is defined within the scope of beneficence towards the participants, even if this effect is indirect, e.g. activities, such as staff training, administrative jobs or publicity, which should all be of advantage to the participants ultimately through improved service provision. Certainly, any active involvement of service users should be underpinned by a clear and perceptible purpose of benefit.

This creates a tension for *research activities* involving service users.[2] Their sense of engaging with service provision is about experiencing support from volunteers and practitioners. By taking part in research, often about sensitive and intimate issues in their lives, participants are supporting researchers, potentially putting their own well-being at stake. Although the research process can be itself a beneficial experience (van Blerk and van Blerk 2015), giving their time, investing in a trust relationship with researchers, accepting the emotional burden of telling and reliving significant personal experiences, with little control over how these will be shared beyond their investment in another person, has the potential to be draining as well as beneficial. While Kye's current situation is with service users who lead as volunteers at the drop-in, which shifts issues of support and volunteer-practitioner-user relations to an extent, these same issues regarding research remain.

In PAR frameworks, in particular, the aim is that research is driven by the participants: they should have greater control over the research questions and designs, and be involved in analysis, and the dissemination of data and findings. In this approach, taking part in the research process is intended to be of benefit to participants in a variety of ways. There is a critical body of work attesting to the complex power relations and messiness of research that is caught up in action research, the co-production of knowledge and 'giving back' to participants (Cahill 2007; Askins and Pain 2011; Shaw 2016). Not least, in the reality of academic work, we pursue other motivations, such as addressing 'wider' questions of relevance to our academic peers, writing for publications in recognised academic outlets, designing methodologies that would be adequate to these aims rather than to the direct interest of participants (e.g. those that incorporate research and learning experience), and meeting requirements of our funders.

Our argument here is a nuanced shift, but that as volunteer-practitioner researchers, we employ strategies to *mitigate the non-beneficence aspect* of any research work. Such strategies/activities certainly overlap with those adopted in many participatory and collaborative forms of research; we highlight here that, in/through all of these, the emphasis is on maintaining relationships and volunteer-practitioner positions *over* advancing research. So, we seek to be fully open and transparent about our motivations and expectations, and explicitly support participants to reflect on their decision to be involved, potentially increasing denial or withdrawal of consent. We seek to minimise activities that are not directly beneficial or risk undermining

---

2   There is also a tension here regarding co-practitioners and volunteers, less directly perhaps: they have service users' benefits as their priority, and their own participation in research should not detract from this.

organisational support and work. Matej, for instance, did not conduct interviews or initiate informal conversations with young participants on themes that were not an element of youth work praxis, though which would have been relevant to research. Kye has put on hold previously agreed plans to train community researchers, as people initially signed up to this experience shifted in their support needs.

Overall, then, we seek to be attentive to the motivations and interests of participants to engage with us in research activities, aware that regardless of how important we believe our research to be, and how we envisage it to have potential benefits for them, their reasons to take part are often about enjoying the research activity or participating because they wish to give (something) back to us – and we should only proceed if such motivations do not jeopardise ongoing support work. We try to develop outputs with participants, and invite their comments, again cognisant that any lack of interest and time is commonly due to other needs, and finding ways to work with and around this where relevant and possible. This has lead, in several instances beyond the cases we present here, to not writing academically from research projects.

So far, we have illustrated that the notion of research ethics cannot be assessed within pre-formulated codes of conduct and has to be recognised in a dynamic and embodied co-presence with others, reiterating previous work (cited above). Further, we point here to *the importance of the relationships* we develop and maintain with research participants/service users. In the following two sections, we first reflect on the character of these relationships and their implications for research ethics, and then seek to re-position these relationships within the wider context of academic praxis.

Thinking through relationships: friendship-liness

Despite the language of service and benefit 'provision', relationships between practitioners and service users are not one-directional. Practitioners are certainly not the only active agents in such relationships and service users are not passive and dependent recipients. As with other care settings (Bowlby et al. 2010), the helping and caring professions are grounded in mutuality between those involved in either role, and such interdependent relationships are at the heart of care experiences (Cloutier et al. 2015). Service users' contribution to the relationship, their (often tacit) care, compassion, time, trust, sharing and embodied partaking in contract and boundary negotiations are all crucial for the day-to-day evolving of ties that hold the service and, in our experience, research together. Additionally, service users might contribute various extents to the organisation management, setting its priorities through involvement in steering committees/user groups, re/designing codes of conduct or directly participating in volunteering roles.

As researchers *and* volunteer-practitioners, we are trained and prompted to reflect on our relationships with others in the field. Thinking through mutuality and interdependence raises questions regarding how our relationships are perceived and understood by research participants. We have found that the ongoing, regular, situated, intense, often spontaneous and emotionally charged character of volunteer-practitioner research praxis predominantly leads to intimate relationships, which often approximate to friendships. Both of us have been called and considered as a friend by research participants, in the cases drawn on in this chapter and other projects.

Friendships are described as enacted through everyday spatial practices and connecting to wider social, cultural and political relations and processes, being an important form of intimacy in increasingly mobile and interconnected geographies (Bunnell et al. 2012). Volunteering week-in-week-out, developing a certain kind of 'working relationship' that is often intimate in a variety of ways, can certainly lead to relationship-building in research attaining a status of friendship – and therein lie many ethical questions. Normatively, friendships are understood and constructed as initially prompted through something in common (shared school, work, community, hobby, activity, and so on), with bonds further strengthened through shared spatial, material and emotional experiences. Friendships in research and professional praxis are often not prompted thus; they do not emerge through organic meetings enabled through similarities, rather the 'getting together' is effortful, political, a will to engage with people as much different than having something in common (de Leeuw et al. 2012). Yet bonds are strengthened through shared spatial and emotional experiences. And while material differences largely remain (in terms of socio-economic positions), other intersectional physicalities and socio-cultural positions may be shared.

We thus argue that we should pay attention to *friendship-liness*, as a lens through which we can partially make sense of our relationships with research participants in the volunteer-practitioner context, recognising those limits and tensions caught up in doing so. Much of our experience in volunteer-practitioner work resonates with our understanding of friendship, enlivened in moments and continuities of care, compassion, conversations and a variety of big and small gestures shared with participants. This particularly includes relationships with volunteer and staff colleagues, in which we could assume lesser asymmetry in power relations (due to more similar class and education positions), and which seem on the surface less problematic for ethics of research. Yet, as de Leeuw et al. (2012) point out, such asymmetries can never be annihilated and they create, among other issues, the risk that participants who become friends with researchers may feel reduced capacity to express concerns about the research process, particularly if researchers claim or aspire to have overcome inequalities.

Indeed, different expectations and permutations of trust and understanding will always impact on the relationships that we as volunteer-practitioner researchers establish with other practitioners. For instance, when for academic work-related reasons Kye missed the drop-in Christmas party (after being involved in preparations), reactions from staff and volunteers ranged from being upset for her (that she missed the event), feeling more broadly disappointed that she could not come (thinking that everyone should attend for a 'successful' party), to experiencing a break of trust (suspicion regarding her commitment). Although a colleague to all and friend to some, Kye's researcher identity created a certain form of tension and isolation, as this absence placed her apart from others.

Relationships with service users are even more difficult to conceptualise as friendships, despite often revealing signs of such relations. In Matej's work with children and young people, his community organisation had an ethical code that prohibited relationships outside the scope of youth work, including a ban on sharing personal details, phone number or communicating via social media. Not least for his participants' age, Matej would hardly conceive of his relationships particularly with younger children as friendships, but some of the children referred to him as a friend. Matej's challenge was not as much to resist or accept such a label but rather seek to understand what the children understood by inviting him to be their friend – often a label for non-judgmentality and acceptance, protection from violence in their lives, and participation in play, fun and conversations. Their conceptions of friendship differed from Matej's but they closely corresponded with what he perceived his youth worker role to be, and in turn informed his ethical presence as a researcher.

Key here is that, as volunteer-practitioner researchers, we experience, value and nurture moments that are *resonant of friendship* – the coalescence of trust, respect, care, compassion and close sharing of the emotional and personal. Yet, we should not easily identify our relationships with research participants as friendships; rather, these *elements* of friendship are complicated and challenged by patterns of difference that we bring both as academics with research motivations, and also as social subjects – aged, gendered, classed and racialised. Additionally, they are enveloped within participants' expectations and understandings of the role of volunteer-practitioner as someone who is 'there for them' and in our commitment to act beneficially, but within set boundaries (of time, space and interactions).

Seeking to describe this form of relationship, we suggest friendship-liness to capture such imperfect yet intense, situational yet grounded, and powerful yet fragile bonds with people we meet through action research. Friendship-liness refers to more than friendliness; it is more than kindness, pleasantness, openness and approachability. We understand it as a relationship of care resembling friendship that is situated in asymmetric positions through which we navigate in the pursuit of particular (research) and broader (social transformation) goals, enacted in day-to-day care praxis as volunteer-practitioners. It refers to an ethical approach to relationships that, building upon ethics of care, approximates friendships but proceeds within partially deliberate geographies and temporalities centred round our academic work.

Distinct from a 'usual' friendship, then, friendship-liness conceives the very particular kinds of relationship we may develop in research. As an ethical *approach*, it is also intended to convey such relationship-building as a *process*, never concretely achieved: emotional labour is always required to develop and maintain such relations, and there is a politics of care required. Indeed, in writing this chapter, we discussed the diversions in our experience, which should be made transparent as all research processes are situational and positioned in complex and unique ways. For Kye, friendship-liness has involved *becoming friends* with some fellow volunteers, meeting outside the drop-in and research spaces. Such becoming remains unstable, and while morphing into more 'usual' friendship, perhaps, always complicates and is complicated by her researcher identity. For Matej, the process of friendship-liness had to also hold specific boundaries in place with young people. He too experienced moments of becoming friends with children and young people – often tacit, unspoken and only subsequently recognised. Yet, more than Kye's, his experience is also one of *unbecoming* friends – due to substantive, primarily age-related differences between him and the children, and due to his conscious efforts in adhering to the organisational code of practice, which prevented him from sharing details about his life and restricted his contact with the children to the space and time of youth work. Aware of such differences, in the following section, we seek to outline how the praxis of friendship-liness weaves beyond individual relationships with co-practitioners, volunteers and service users.

Re-enacting and re-placing relationships

As Raghuram (2016: 511) exhorts "care ethics needs emplacing", it's theorisation must be extended to take account of the diversity of how ethical practices and concepts are constitutive of relationships in socially and culturally specific circumstances. This means recognising and exploring various types and spaces of multicultural ethics globally, but we also need to consider relations between multiple kinds of care and ethics as they intersect within localities, and the relationships that develop. Whatmore (1997) proposes that along with situating relational acts, key to developing a relational perspective on ethics lies in "displacing the fixed and bounded contours of ethical community" (50). We understand this as a challenge to critically reflect on who we are interconnected within research, and what it means for questions of ethics. That is, how our research praxis is extended in directions that exceed individual relationships, and also folds back into processes of friendship-liness.

As volunteer-practitioner researchers, we develop relationships with individuals *and* with their communities, families, friends. These relationships might not be as direct and embodied, but these other actors are entrenched in the lives of participants. As such, their presence is vocalised, or often more implicitly *felt*, in our interactions with service users, volunteers and practitioners, and they are entangled in the ongoing relations and practices of care that underpin service provision and research praxis. Echoing the literature on care at-distance (Mason and Whitehead 2012) and on networking and intermediaries in care practices (Blazek et al. 2015), we experience volunteer-practitioner research settled in *complex webs of relations*. Thus, what we say and do travels across geographies, from the volunteer-practitioner and research spaces to multiple others, and may have affects and effects we cannot know but should consider in terms of ethics and care. Likewise, otherly emplaced care ethics work across geographies to influence multiple relationships within organisational and research spaces.

So while the potential of social researchers' public contribution ('doing good') is often seen to come from research skills and competence (cf. Mitchell 2008), we suggest that this potential should be more broadly perceived. Echoing literature that points to the need to be collaborative by being attentive and responsive to the realities of participants and community partners (PyGyRg 2012), beyond a participatory process of determining research design with participants, we foreground that academics can/do bring more than research skills and professional competence to research relationships. For example, we operate across different networks that can be brought together, facilitating shifts in thinking and action. A key 'achievement' of Matej's knowledge exchange grant (attached to his PhD research) was putting in touch youth work organisations from the UK and Slovakia, who established international collaborations without him being involved any further. Kye has drawn on contacts in Amnesty International, developed through long-term personal involvement, for a workshop in the drop-in raising awareness of asylum-seeking.

Not least, diverse, 'small' practices can contribute in varied and messy ways through such webs of spatialised relations towards broader societal shifts. Such change is often an ambition of research: to inform policy, to support transformative action with knowledge and interpretation. Yet, collaborating closely with partner organisations as volunteer-practitioner researchers, we find ourselves entangled in a fluid multiplicity of roles. Being community advocates in academic writing and engagement with policy-makers is certainly one position. Between us, we also have experience of identifying (non-research) funding opportunities and writing funding bids, cleaning premises and repairing broken furniture, editing a website and driving the minibus on a residential trip, being trainees and trainers of practitioner colleagues, and administrators contributing to the running of organisations with clerical work. We also bring commitment and – importantly – often time and resources that are unavailable to community-based organisations, particularly in times of austerity. Moreover, while we might seek to give something extra[3] in line with our motivations and a politics of care, we find that such involvement and activities are simultaneously demanded of us, precisely through the relationship-building that we outline above: we take on certain responsibilities that are asked of us.

---

3 'Giving extra' points towards understanding research roles and relationships as situated in these wider webs of relations, rather than a narrow two-way giving-to and giving-back.

Further, if we are enmeshed in wider webs of participants' relations, then so too are they – and our experiences as volunteer-practitioners – enmeshed in our social and professional lives. We should be attentive to what (extra) we receive through research and can take elsewhere along our academic and non-academic journeys. Echoing Williams' (2016) reflections on research volunteering, it is crucial to note that caring relationships with those outside volunteer-practitioner research settings, such as with students and colleagues, are also mutually co-constitutive of relationships in research and beyond work environments. Volunteer-practitioner research is based on searching out and investing in certain kinds of relationships, performed through particular acts, many of which can be relocated and reframed in different social contexts.

One significant aspect for both of us is that working as volunteer-practitioners has helped us re-develop listening skills in new ways, such as attending to body language and non-verbal communication, through organisational training, and through experiential learning day-to-day, as well as from our co-practitioners and volunteers. This then permeates into departmental settings, student supervision meetings, seminars and lectures. Training in 'soft' or social skills is a standard practice in voluntary sector organisations, yet often glossed over in research methods handbooks, ethics texts or indeed supervisory practices (Shaw, 2016). More broadly, the emphasis in the helping and caring professions *on relationships as central* to service provision and progressive change, calls into question for us how we re-enact and re-place relationships across geographies of work and home … not least in terms of ethical praxis and politics of care beyond research, but in everyday academic endeavours … and not least this has been the inspiration underpinning this chapter.

Rather than end by way of a conclusion, then, we would stay open to notions of friendship-liness as ethical commitment and process, by joining our voices to recent and increasing calls to inculcate more caring ways of being academics, in response to "deepening neoliberal logics that turn the academy into a place where care and love become radical acts of resistance and transformation" (Lopez and Gillespie 2016: 1689). While this chapter has emphasised the ethical dimensions of a particular framing of research relationships, as volunteer-practitioners, we invite readers to consider *the role of relationships as central to fieldwork praxis across diverse methodologies, and across diverse academic roles and positions*. Friendship-liness as a lens through which we can partially make sense of relationships with each other, that includes valuing each others' skills and experiences brought into departments from research, recognising the limits and tensions in doing so, yet enlivened in moments and continuities of care, compassion, conversations and a variety of big and small gestures shared.

What we ultimately bring into this writing space, then, comes in part through the extra we experience in research: foregrounding friendship-liness as an ethical approach, as process, as a politic of care.

References

Askins, K. (2007) 'Codes, committees and other such conundrums!', *ACME: An International Journal for Critical Geographies*, 6 (3): 350–359.

Askins, K. (2016) 'Emotional citizenry: Everyday geographies of befriending, belonging and intercultural encounter', Transactions of the Institute of British Geographers, *41* (4): 515–527.

Askins, K. and Blazek, M. (2017) 'Feeling our way: Academia, emotions and a politics of care', *Social & Cultural Geography*, *18* (8): 1086–1105.

Askins, K. and Pain, R. (2011) 'Contact zones: Participation, materiality and the messiness of interaction', *Environment & Planning D: Society & Space*, *29* (5): 803–821.

Bignante, E., Mistry, J., Berardi, A. and Tschirhart, C. (2016) 'Feeling and acting 'different' emotions and shifting self-perceptions whilst facilitating a participatory video process', *Emotion, Space and Society*, *21*: 5–12.

Blazek, M. (2014) 'Migration, vulnerability and the complexity of violence: Experiences of documented non-EU migrants in Slovakia', *Geoforum*, *56*: 101–112.

Blazek, M. (2015) *Rematerialising Children's Agency: Everyday Practices in a Postsocialist Estate*. Bristol: Policy Press.

Blazek, M., Smith, F. M., Lemešová, M. and Hricová, P. (2015) 'Ethics of care across professional and everyday positionalities: The (un)expected impacts of participatory video with young female carers in Slovakia', *Geoforum*, *61*: 45–55.

Bond, T. (2015) *Standards and ethics for counselling in action*. London: Sage.

Bowlby, S. R., McKie, L., Gregory, S. and MacPherson, I. (2010) *Interdependency and care over the lifecourse*. Oxford, Routledge.

Bunnell, T., Yea, S., Peake, L., Skelton, T. and Smith, M. (2012) 'Geographies of friendships', *Progress in Human Geography*, *36* (4): 490–507.

Cahill, C. (2007) 'Repositioning ethical commitments: Participatory action research as a relational praxis of social change', *ACME: An International Journal for Critical Geographies*, 6 (3): 360–373.

Chatterton, P., Fuller, D. and Routledge, P. (2007) 'Relating action to activism: Theoretical and methodological reflections', in Pain, R. and Kesby, M. (eds) *Connecting people, participation and place*. London: Routledge.

Cheung Judge, R. (2016) *Transformational journeys: Volunteer tourism, non-elite youth and the politics of the self*. Unpublished PhD thesis. London: University College London.

Cloutier, D.S., Martin-Matthews, A., Byrne, K., Wolse, F. (2015) 'The space between: using 'relational ethics' and 'relational space' to explore relationship-building between care providers and care recipients in the home space', *Social & Cultural Geography*, *16* (7): 764–782.

Conradson, D. (2003) 'Spaces of care in the city: The place of a community drop-in centre', *Social & Cultural Geography*, *4* (4): 507–525.

Corey, M.S. and Corey, G. (2014) *Becoming a helper*. Stamford: Cengage Learning.

Corey, C., Corey, G., Callanan, P. and Corey, M.S. (2014) *Issues and ethics in the helping professions*. Stamford: Cengage Learning.

Cornforth, S. and Claiborne, L.B. (2008) 'When educational supervision meets clinical supervision: What can we learn from the discrepancies?', *British Journal of Guidance and Counselling*, *36* (2): 155–163.

de Leeuw, S., Cameron, E. S. and Greenwood, M. L. (2012) 'Participatory and community-based research, Indigenous geographies, and the spaces of friendship: A critical engagement', *The Canadian Geographer: Le Géographe Canadien,* *56* (2): 180–194.

Denzin, N., K., Lincoln, Y.S. and Tuhiwai Smith, L. (2008) *Handbook of critical and indigenous methodologies.* Los Angeles: Sage.

Dickens, L. (2017) 'World making, critical pedagogies and the geographical imagination: Where youth work meets participatory research', *Antipode, 49* (5): 1285–1305.

Dickens, L. and Butcher, M. (2017) 'Going public? Re-thinking visibility, ethics and recognition through participatory research praxis', *Transactions of the Institute of British Geographers, 41* (4): 528–540.

Dyer, S. and Demeritt, D. (2009) 'Un-ethical review? Why it is wrong to apply the medical model of research governance to human geography', *Progress in Human Geography, 33* (1): 46–64.

Ellis, C. (2007) 'Telling secrets, revealing lives: Relational ethics in research with intimate others', *Qualitative Inquiry, 13* (1): 3–29.

Evans, R. (2016) 'Achieving and evidencing research 'impact'? Tensions and dilemmas from an ethic of care perspective', *Area, 48* (2): 213–221.

Fuller, D. and Askins, K. (2010) 'Public geographies II: Being organic', *Progress in Human Geography, 34* (5): 654–667.

Gilligan, C. (1982) *In a different voice: Psychological theory and women's development.* Cambridge: Harvard University Press.

Hall, S. M. (2009) "Private life' and 'work life': Difficulties and dilemmas when making and maintaining friendships with ethnographic participants', *Area, 41* (3): 263–272.

Klocker, N. (2015) 'Participatory action research: The distress of (not) making a difference', *Emotion, Space & Society, 17*: 37–44.

Kovach, M. (2010) *Indigenous methodologies: Characteristics, conversations and contexts.* Toronto: University of Toronto Press.

Lopez, P. and Gillespie, K. (2016) 'A love story: For 'buddy system' research in the academy', *Gender, Place & Culture, 23* (12): 1689–1700.

Malam, L. (2004) 'Embodiment and sexuality in cross-cultural research', *Australian Geographer, 35* (2): 177–183.

Mason, K. (2015) 'Participatory action research: Coproduction, governance and care', *Geography Compass, 9* (9): 497–507.

Mason, K. and Whitehead, M. (2012) 'Transition urbanism and the contested politics of ethical place making', *Antipode, 44* (2): 493–516.

Mitchell, K. (ed) (2008) *Practicing public scholarship: Experiences and possibilities beyond the academy.* Oxford: Blackwell Publications.

Pain, R., Kesby, M. and Askins, K. (2011) 'Geographies of impact: Power, participation and potential', *Area, 43* (2): 183–188.

Proudfoot, J. (2015) 'Anxiety and phantasy in the field: The position of the unconscious in ethnographic research', *Environment and Planning D: Society and Space, 33* (6): 1135–1152.

PyGyRG (2012) 'Connectivity, creativity, hope, and fuller subjectivities: A communifesto for fuller geographies', https://radicalantipode.files.wordpress.com/2012/12/pygyrg-reply.pdf.

Raghuram, P. (2016) 'Locating care ethics beyond the Global North', *ACME: An International Journal for Critical Geographies, 15* (3): 511–533.

Richardson, M.J. (2015) 'Theatre as safe space? Performing intergenerational narratives with men of Irish descent', *Social & Cultural Geography, 16* (6): 615–633.

Sevenhuijsen, S. (1998) *Citizenship and the ethics of care.* New York: Routledge.

Shaw, J. (2016) 'Emergent ethics in participatory video: Negotiating the inherent tensions as group processes evolve', *Area, 48* (4): 419–426.

Skovholt, T. M. and Trotter-Mathison, M. (2016) *The resilient practitioner: Burnout and compassion, fatigue prevention and self-care strategies for the helping professions.* Abingdon and New York: Routledge.

Smith, F. M., Blazek, M., Brown, D. M. and van Blerk, L. (2016) "It's good but it's not enough': The relational geographies of social policy and youth mentoring interventions', *Social & Cultural Geography, 17* (7): 959–979.

TallBear, K. (2014) 'Standing with and speaking as faith: A feminist-Indigenous approach to inquiry [Research note]', *Journal of Research Practice, 10* (2), N17, http://jrp.icaap.org/index.php/jrp/article/view/405/371.

Taylor, M. (2014) 'Being useful' after the Ivory Tower: Combining research and activism with the Brixton Pound', *Area, 46* (3): 305–312.

Tuhiwai Smith, L. (2012) *Decolonizing methodologies: Research and Indigenous peoples.* London: Zed Books.

van Blerk, L., van Blerk, D. (2015) 'Biographical interviews as emotional encounters in street youth's lives: The role of research in facilitating therapeutic intervention', in M. Blazek and P. Kraftl (eds) *Children's emotions in policy and practice: Mapping and making spaces of childhood.* Basingstoke: Palgrave, pp. 189–203.

von Benzon, N. and van Blerk, L. (2017) 'Research relationships and responsibilities: 'Doing' research with 'vulnerable' participants: Introduction to the special edition', *Social & Cultural Geography, 18* (7): 895–905.

Whatmore, S. (1997) 'Dissecting the autonomous self: Hybrid cartographies for a relational ethics', *Environment and Planning D: Society and Space, 15* (1): 37–53.

Williams, M. J. (2016) 'Justice and care in the city: Uncovering everyday practices through research volunteering', *Area, 48* (4): 513–520.

Zhang, J. J. (2017) 'Research ethics and ethical research: Some observations from the Global South', *Journal of Geography in Higher Education, 41* (1): 147–154.

Part II

Research ethics in the wider academic context

9 Illegal ethnographies

Research ethics beyond the law

Thomas Dekeyser and Bradley Garrett

Prologue: an epistemological deadlock[1]

We were handed a bundle of court documents which included a statement from an independent 'expert' on ethics. The statement alleged that author two had been unethical in his research practices because he had participated in illegal activity.[2] Over the next week, author two and a colleague constructed a 17,000-word response, amassing evidence from the PhD dissertation, academic sources and disciplinary association guidelines on ethical research practices which make clear that researchers can study illegal activities while maintaining an ethical position. Discussing it with legal counsel, it dawned on us that we had reached an epistemological deadlock. The witness and prosecution sought to rationalise ethics as a transcendental code that could be applied equally across all situations, suggesting that the illegal was categorically unethical. Conversely, we marshalled evidence from a less prescribed understanding of ethics that recognised the fluidity, challenges and situatedness of doing ethnographic research. In the midst of these discussions, we were forced to confront the congruences and incongruences between ethics and law. We wondered what it might mean to argue convincingly for situated ethics in the courtroom.

Introduction

> Situate yourselves as close as you can to the perpetrators of crime and deviance, to the victims, to the agents of legal control; put yourselves, as best as you can and for as long as you can, inside their lives, inside the lived moments of deviance and crime. You won't experience it nicely, and if the danger and hurt become too much, be glad of it. Because as

---

1   A previous version of this chapter was previously published in *Area* as 'Ethics does not equal law'.

2   Dr Garrett was charged, along with 11 research participants, with 'conspiracy to commit criminal damage' following a multi-year ethnographic project with urban explorers who 'recreationally trespassed' into off-limits architecture (see Garrett 2013, 2014).

DOI: 10.4324/9780429507366-9

near as you will ever get, you have found your way inside the humanity of crime and deviance.

(Hamm and Ferrell 1998: 270)

Can breaking the law as researchers ever be ethical? Or should we, as researchers, by default curb our dedication to ethical research to the terms set out by local, national and international law? In this chapter, we investigate how, in certain instances, the difference between ethics and law puts the practising of situated ethics to the test. Human geographers are increasingly interested in post-Kantian understandings of ethics as an ongoing, situated, multi-sensual sensibility and set of prompts, rather than as a set of transcendental codes determined in advance of field encounters that can be rationally applied (see for instance England 1994; Rose 1997; Whatmore 1997; Cloke et al. 2000; Smith 2001; Routledge 2002, 2009; Valentine 2005; Horton 2008; Popke 2009; Dyer and Demeritt 2009; Hynes 2013; Dickens and Butcher 2016). Situated ethics is a mode of conduct to move towards and along (Madge 2007); an evolving critical sensibility aimed at cultivating "a disposition and an openness to difference, the multiplicity of life and, crucially, uncertainty" (Braidotti 2006; Bissell 2010: 85). Here, ethics is about debate, dialogue and compromise rather than consensus (Madge 2007). These mediations may take place in a public forum, but also take place, importantly, within ourselves and with our project participants. This process exceeds "the impartial and universal enactment of instrumental reason" exchanged among and within rational, cognitive subjects (Whatmore 1997: 38), to acknowledge the embodied, practised qualities of negotiating ethical agency as unfolding within webs of connectivity between researcher and research participants (Whatmore 1997; Routledge 2002).

Professional institutions, funding bodies and university departments are also recognising the importance of situated, embodied and relational ethics; an ethics that engages ethics not as "an attribute of a pre-existing ethical subject but as a potential mobilised within particular creative instances" of spatio-temporal unfolding (Hynes 2013: 1931). The American Association of Geographers (2009: np), for instance, acknowledges "[n]o one statement can possibly cover the range of ethical matters confronted by geographers." Likewise, one geography department in the UK encouraged geographers to "[r]emember that there is no blueprint for ethical practice." Yet, despite geographers' and geography's ostensible embrace of situated ethics, the law seems to create parameters human geographers are unwilling to cross. The result is a simple equation of the lawful and the ethical, indifferent to the various ways lawful behaviour may sometimes emerge as unethical, and ethical activity as unlawful.

The equation between ethics and law pushes ethnographic work concerned with explicit embodied interaction with what lies beyond legal borders into the realm of taboo. It forces institutions to back down on their statements of situated ethical practices in order to comply with the law, and can even put researchers and institutions in situations where they are expected to act unethically with respect to research participants. In this chapter, we explore the assumption that ethics and law are necessarily and unavoidably in alignment and how this assumption may produce strict, static ethical arrangements that inhibit the 'doing' of situated ethics. Given geographers are increasingly engaging the illegal (Hall 2012) – including the geographies of burglary (Bernasco et al. 2015), illegal occupations (Vasudevan 2015), drug dealing (Allen 2005), street racing (Falconer and Kingham 2007), counterfeit products (Gregson and Crang 2017), illegal abortion provision (Calkin 2019), maritime piracy (Hasting 2009), graffiti (McAuliffe and Iveson 2011), urban exploration (Garrett 2013) and subvertising[3] (Dekeyser 2021) – these questions must have a role in our disciplinary conversations. When dealing with illegality, there is arguably even greater potential for both researcher and research participants to experience malevolent effects of unethical research.

Human geographers are only beginning to grapple with these issues, but disciplines such as Criminology, Sociology and Anthropology have a longer history of debating both precedent case studies and conceptual scenarios. For instance, 'edge ethnography,' in which the researcher engages "in behavior that is physically, legally, socially, or otherwise dangerous," is a recognised and widely-discussed methodology in criminology.[4] Likewise, the *American Anthropological Association* (AAA) includes among its teaching resources case studies where researchers are forced to confront mismatches of law and ethics.[5] Importantly, the resources refuse final judgements or statements of how to act and, instead, repeatedly note that ethics are messy, complicated and relational.

It is in this spirit that we, in this chapter, consider our own and others' experiences with illegal ethnographies to raise a series of questions. Rather than casting judgements or proposing a 'framing' for illegal ethnographies, we investigate situations in the stages preceding, during and after fieldwork where our speculative embrace of situated ethics runs up against the law, or where an ethical evaluation of the complex relations between researcher, research participants and research institutions that make up a particular ethnographic situation calls for us to break the law to remain ethical. We interrogate the possibility of a more situated approach capable of providing support to researcher and research participants that facilitates ethnographic entanglements responsibly, promoting greater and contingent interactions with ethical concerns. We start, then, where all ethnographic projects do: with their inception.

---

3 Subvertising is the illegal act of materially intervening into outdoor advertising spaces through methods of replacement, removal, supplementation, reversal or destruction.
4 See http://criminal-justice.iresearchnet.com/criminology/research-methods/edge-ethnography/ (accessed 27 April 2017).
5 See https://www.americananthro.org/LearnAndTeach/Content.aspx?ItemNumber=12912 &RDtoken=38123&amp&navItemNumber=731;userID=5089 (accessed 39 April 2019).

Inception

> Yesterday's PhD interview started off seamlessly. We shared theoretical
> interests, political hopes and I think I responded adequately to some of
> the potential supervisor's more provocative questions. Towards the end,
> however, when we started discussing the ethnographic component of the
> proposed methodology, the temperature of our conversation dropped.
> 'Hmmmmm, this all sounds a bit too innovative for this university,' the
> proposed supervisor said. It felt like this signalled the end of his, and
> his university's, interest in my project with subvertising practitioners.
>
> (Author one, MA research diary, January 2015)

Oftentimes, confronted with fieldwork that veers towards legal boundaries,
many of us sitting on ethics committees or working as supervisors might
have heard, thought, or even said, "that will never get through the ethics process." A central, *a priori* application of codes remains prevalent in
the process of research inception. Horton (2008: 367) suggests that one of
Geography's central guidelines is that, as researchers, our "everyday practices must comply with national legislation, local rules and regulations."
This line-guarding is found within numerous university departments, who
emphasise the obligation for researchers to comply with local, national and
international legal frameworks. But what are the ethics of patrolling what
we can do, of performing mechanical methodological restriction through
uncritical reliance on the ethical coding framed by local, national and international law?

It is worth turning to the scholarly field of criminology, and more particularly, to the methodological emergence of 'edge ethnographies' for guidance
(Hamm and Ferrell 1998; Miller and Tewksbury 2010). As Ferrell and Hamm
(1998: 34) write, if "the law in part reflects, incorporates, and perpetuates
social privilege and social injustice," then a dedication to the lawful may, in
some cases, debilitate an orientation towards ethical engagement with field
encounters. Research practices critical about, and possibly even confrontational towards, legal systems, can for that reason be more ethical, where
they seek correct prejudice, injustice, and oppression (see Caitlin et al. 2007).
In this vein, some geographers have cultivated collaborations with activists
and activist organisations, aligning shared goals through various modes of
nonviolent direct action (e.g. Nast 1994; Fuller 1999; Routledge 2002) that
skirt the edges of or move beyond the realm of the legal.[6] Following Hannah
Arendt, who emphasised the "not altogether happy theoretical marriage of
morality and legality, conscience and the law of the land," (1972: 52) and a

---

6   Illicit or legally dubious forms of nonviolent direct action, undertaken by geographers,
    range from performing a fake identity (Routledge 2002) to protesting in illicit places (Lee
    2013) and blockading power plants.

long history of scholarly civil disobedience (Thoreau 2001), such research underscores how challenging laws where they perpetuate injustice is not only potentially ethically acceptable but also *necessary*.

The geographer Paul Chatterton, for example, highlighted in a research-er's defence statement to a UK jury that his occupation of a coal train in Yorkshire with research participants was justified and imperative in light of the coal-fired power station's 'deadly and urgent threat to society' respon-sible for '180 deaths a year' (see Wainwright 2009). While the actions of the researcher may have been technically illegal in that moment, the broader awareness of law and the particular contexts of the action here rendered the illegal actions ethical and even potentially legal under 'lawful excuse,' where an unlawful act takes place to prevent more serious harm. Had Chatterton been prevented from undertaking this research by his institution, none of the knowledge gained would be available for debate. Refraining a priori from researching illegal spaces and practices, through relying on the famil-iar mantra to 'play it safe' and 'not choose a topic that might lead you to break the law,' (Denscombe 2012: 168) might therefore be more dangerous and unethical than cultivating a situated ethics through the messy encoun-ters of illegal performances, embodied situated selves, community ethics, legal systems, ethics committees, research aims and the need to protect research participants.

The criminologist Jeff Ferrell, arrested while painting trains with graffiti writers, argues that illegal ethnographies might reveal "part of the social world that remains hidden by more traditional techniques" (1998: 25). Writing of his ethnographies, Ferrell notes that knowledge comes to us as much in "the pits of our stomachs, in cold sweats and frightened shivers, as in our heads" (1998: 30). Only through direct immersion is it possible to vis-cerally trace the unlawfulness of an unlawful practice, to share with project participants the adrenaline-fuelled anxieties central to particular kinds of illegal practice. If we agree that it is sometimes across the legal bounda-ries of social reality that important practices and the promise of alternative futures lies (for instance, in squatting communities, pirate radio networks or illegal favela settlements), and if we consider these practices and futures as embodied, imagined, performative and affective (perhaps never emerging in the public realm) *and* as semiotic and discursive (through their pub-lic mediations and media productions), then direct immersion into illegal practices is not only ethically acceptable but also critical to sociopolitical apprehension. In other words, if worlds are built both legally and illegally, our research into those worlds cannot be consigned to one side of that fence.

We want to emphasise that ethical negotiations do not start from, nor end with, 'the field' (Till 2001). And we ask: should the scales of knowledge lost or gained through ethnographies into the unlawful not be given as much consideration as any other mode of research conception by departments and ethics committees? For we cannot know, in advance, what forms of affective and ethical potential ethnographic investigations into the dark web, drug use or far-right extremism may unveil.

In the field

> When I arrived at the subvertising workshop a few hours ago, people were already getting seated, notepads and information packs in hand. The community centre was packed. Two men arrived late, just after I started my presentation on subvertising history and theory. They slotted in the background and kept to themselves, didn't mingle with any other participants. I didn't think much of it until later, when during the practical part of the workshop, they spoke for the first time. *Are you thinking of doing any digital billboard takeovers?* Slightly uncomfortable, my co-presenter Donna responded after a moment of silence: 'No, no, these are legally a wholly different thing. What we're doing is closer to criminal damage ...' As people gathered after the workshop, the two same men – perhaps ten to twenty years older than the average crowd – approached us with a certain sense of urgency. *Can we have your full name?* Donna, by now severely suspicious, responded: 'Erm, no, I don't see why you'd want that'. *We want to follow you on Facebook.* 'Oh, okay, well you can follow the collective's name,' Donna suggested. They kept pressing until we pretty much ran away from them.
>
> Not knowing what to do, I rang my friend Amelia and talked her through the details: the layout of the evening, the men's looks, their not-so-subtle questioning, how they had wrenched as much info as possible, and how they had this peculiar determinacy and almost arrogance in their voices.
>
> 'Private investigators', she responded shortly. And added, before I had taken in what that actually meant, 'prepare for a raid'. Now I'm sitting here, in my living room. I'm actually terrified. FUCK. I guess this is what my research-participants have had to deal with all along. It now makes sense, more viscerally than ever.
>
> (Author one, field notes, September 2016)[7]

This anecdote underscores a point that arises in every ethnographic project where fieldwork becomes excessive to expectations. In this vignette, however, this excessiveness stems from a closeness with project participants that requires negotiating legal lines. An educational workshop became, through the mere presence of private investigators, a risky scene, with potential legal implications for both research participants and researcher. It altered the nature and direction of the event, the researcher becoming, at best, a node holding valuable information for police authorities, and at worst, a

---

7   Pseudonyms are used throughout this chapter to protect research participant anonymity.

criminal accomplice or accessory. As Mitchell (2001: 208) highlights, the role researchers "play in the field and beyond are not strictly and exclusively of their own choosing." It is hard to know, in advance, how a situation will unfurl and on what side of legal boundaries it will land. How does an abrupt change in events, such as the one described, alter the researcher's relation to the moment? How does one act most ethically in a situation that has slipped beyond legal neutrality?

But the question of the relation between situated ethics and law extends beyond the unknowing transgression into legally dubious terrain while in the field – there are times when intentionally breaking the law becomes the most ethical action to undertake. Consider, for instance, a moment when while on a walk a research participant announces a plan to undertake criminal trespass onto a railway, or even just does it without warning. As an ethnographic researcher, aware of the dangers of a solo trespass onto a terrain replete with serious hazards, we might decide it could well be more ethical to participate consciously in the act (if only to look after one another) than to let the research participant depart on a solo journey. The standardised refusal to engage with the unlawful could, in this instance, put a research participant in danger, with potentially fatal consequences.

The positionality of the researcher, in this instance, further complicates the decision to trespass. Routledge reflects on the complex ethical process involved in undertaking illegal trespass (among other legally dubious actions) in the context of political resistance against tourism development in India, where his privileged "whiteness [...] might have vitiated against possible legal sanctions that may have accrued to an Indian person conducting a similar action" (2002: 485), which, alongside gendered and physical qualities, allowed his enactment of the legally dubious to be "[l]aced with a frisson of excitement" (2002: 483) rather than with dread or anxiety.[8] The researcher's positionality, both how they perceive themselves and how they expect to be perceived by others, affects the ethics of deciding to follow trespassers into illegal territory.

These scenarios outline complex situations that defy any assumption "that ethical guidelines and rules can be neatly, rationally applied to a given situation" (Horton 2008: 368). They point at ethical considerations in which an event's generativity itself becomes the 'framework' for making decisions ethically. We can of course never know 'ourselves at a 'transparently conscious' level, at every twist and turn of 'being' (White and Bailey 2004: 138), how the events will unfold exactly and along which paths the

8   As such, Routledge encourages us to reflect on the potentially gendered, racialised, ableist and class-based processes involved in undertaking (and arguably also in desiring to undertake) ethnographies around illegal activity. However, Routledge also emphasises that "various forms of (dramatic, arrestable) direct action have been undertaken in different contexts by people of differing gender, ableist, class, and ethnic backgrounds" (2002: 485).

implications of our own decisions will travel (Katz 1994). Because we can never fully grasp the field's generativity, particularly when decisions have to be made rapidly or at a moment when we feel insufficiently informed,[9] our situated decision-making is bound to fail (Horton 2008). This should not, however, lead us to 'anything goes' scenarios or back towards a premeditated restriction of one's ethnographic activity to legal domains. Rather, we suggest the cultivation of sensitivities centred around "preparation in the sense of planning a mode of attack, a style, a form of engagement" (Gilson 2011: 81) that is capable of tracing "the differential relations, intensities, and singularities that haunt a collective in a moment of perplexity proper to a situation" (Bryant 2011: 41). Carefully reading up on the law and your legal rights, investigating where to find legal support if arrested, working through fictionalised 'ethical dilemmas',[10] writing a reflective research diary, and debriefing with research-participants, colleagues and research institutions (Bashir 2018) can all help craft and re-craft our singular 'mode of attack'. This (re-)crafting requires time, and perhaps even a fixed timing set out in advance to ensure ongoing evaluation.

The development towards immanent evaluation prefigures and transforms along with fieldwork activity. It is informed by an ongoing feedback loop of engagement with our field milieus, research participants, ethics committees, supervisors, colleagues and personal reflections. To task oneself with an unfolding ethics encompassing a multitude of dimensions (of self, participant community, institution, state) can be more tedious or harrowing than a simple equation of the lawful and ethical. Yet, it is perhaps precisely the nonlinear messiness of such a situated ethics that is required for dealing with the ethical complexities of the field.

'After' the field

> I have been charged with 'conspiracy to commit criminal damage' and released on bail. One of the bail conditions, given the alleged 'conspiracy' between myself and my project participants, is that none of us can contact each other. Now I find myself writing an article about my time in the field with them and unable to ask them to approve quotes, check details, or ask for their advice or guidance. In preventing contact with my research participants, now named as co-defendants, I am unable to close this research project in an ethical way, which would demand continued consultation with the community on research outputs.
>
> (Author two, defence document prepared for solicitor, April 2014)

---

9   As Markey et al. (2010) note, it takes time to build relationships, to become respectful to research contexts and participants, in ways that allow for the performance of an ethical research practice.

10 See note 4.

After building relations between researcher and community, there is often no easy 'after the field,' with the field refusing to be spatially and temporally fixed, 'over there,' or back in time (Till 2001). Ethical considerations thus commonly stretch beyond the final day 'in the field.' Like those emerging during fieldwork, not every ethical eventuality after the field can be planned for, asking us to take seriously a post-fieldwork *situated* ethics in at least two ways.

First, the question of participant confidentiality is of particular relevance to ethnographies of illegal practices. In author two's case it was several years after fieldwork concluded that his data – both in raw form on hard drives and printed texts, and in processed form in a thesis, articles and blog posts – became evidence. There is a history of arrested ethnographers being asked and even forced to foreclose personal data on research participants involved in illegal practices (see Hamm and Ferrell 1998). While researchers can protect their data (fieldwork diary, emails, text messages, images, video) through, for instance, encryption of their emails, phones, hard drives, laptops and back-ups 'in the cloud,'[11] this security measurement reaches its limits when the researcher is ordered to reveal their passwords by legal authorities. At this point, the researcher and their host institution face a vital question: is the decision to succumb to pressure from authorities always the most ethical choice?

In 2009, the AAA and the University of Minnesota supported the graduate student sociologist Scott DeMuth, who was jailed for six months for refusing to give up the names of the radical animal rights groups he worked with.[12] Research participant confidentiality was argued to carry greater ethical weight than codes of legality predicated on ideals of (corporate) private property and economic wealth. Adding to this complexity, it is not always easy to ensure confidentiality or anonymity even where access is not granted. Not all communities desire confidentiality, and some will go to considerable lengths not only to decline anonymity but also to destroy any protective cover accorded them (Svalastog and Eriksson 2010). Rather than imposing any clear-cut code of confidentiality practice, the AAA website suggests the ethical researcher will determine what to do in the specific situation.[13]

Second, for Demuth, as for author two, retention of data was also an ethical issue. Data retention has long been a topic of debate. For many years, destruction of ethnographic fieldnotes post-publication was considered standard practice and even demanded by *Institutional Review Board* (IRB) committees in the United States. More recently, the decision has been left to

---

11   It is worth noting that universities, at least in the case of both authors, fail to live up to the necessary contemporary data security demands of doing sensitive ethnographic research. While universities are commonly able to deliver basic data management tutorials, classes regarding encryption of personal hardware, software, online activity and communication with research-participants often have to be learned elsewhere.

12   See https://www.insidehighered.com/news/2009/12/04/demuth (accessed 26 April 2017).

13   See note 5.

the researcher, or increasingly to research councils, some of which require not only assurances of data retention but also its widespread accessibility. For instance, the UK *Economic and Social Research Council* (ESRC), while encouraging the protection of sensitive materials and the need for anonymity, suggests that data must be "findable, accessible, interoperable, and reusable" (Economic and Social Research Council 2015: 2).

In 2015, the anthropologist Alice Goffman, whose research provoked insight into inner city (and often criminal) life in a lower-income neighbourhood in West Philadelphia, revealed that "she shredded all of her notebooks and disposed of the hard drive that contained all of her files out of fear that she could be subpoenaed and thereby forced to incriminate her subjects" (Neyfakh 2015: np). Though extra-disciplinary agents decried this move as unethical and even illegal, Goffman's actions accorded with AAA guidelines which suggest *de facto* destruction of field data unless a compelling (research-led) reason exists for retention. This calls into question 'common sense' policies of data retention. Similarly, an ethics case study in the AAA handbook explores the dilemma faced when a researcher witnesses a murder. When regional police come to interview her, should she turn her research into evidence, destroy her field notes, or, when questioned, plead ignorance? The anthropologist decided to hide her fieldnotes and lie to the police. The website does not pass judgement on the decision, but rather advises that we debate data retention from the outset of our project within its social and cultural contexts.

In line with cultivating situated sensitivity for in-field decision-making, both post-fieldwork scenarios point at the ethical potential of ongoing circuits of communication between researcher, research participants and ethics committees. It is clear that in considering the relationship between law and ethics, we must recognise that the law itself can become a violent tool and that blanket 'rules' leading to categorical statements and a shirking of ethical responsibility under the pressure of legal scrutiny can actually be an unethical position. As Cloke et al. note, "to hide behind ethical standards so as to obscure the real-time dilemmas of research" constitutes an unethical way to engage ethnographic work (2000: 251; see also Dyer and Demeritt 2009: 60). Rather, as understood by the AAA and other major ethics bodies, ethics is composed of a series of complex and situated particularities of individual pieces of research that works less with 'standards' and more with 'specifics,' and doesn't end at the point of publication.

Conclusion

In this chapter, we suggested that an understanding of the *situatedness* of ethics demands careful consideration of the often uncomfortable relationship between the legal and the ethical. We interrogated this relationship before, during and after ethnographic 'field' engagements. In doing so, we suggested that the inherent excessiveness and particularity of field encounters usefully problematises transcendental ethical judgement *prior to* the occurrence and recurrence of research events. We have endeavoured to prompt exploration of what a situated ethics might look like for ethnographies of the illegal, where the equalisation of ethics and legality fails, and suggested that a situated orientation towards unlawful ethnographic fieldwork requires a receptiveness to excessiveness and process that calls into question the stemming of certain kinds of research, the policing of the actions of the researcher or the mobilisation of research data as 'evidence.'

The practice of illegal ethnographies forces reflection on the practising of situated ethics, where at the minimum the law is challenged and at times may even be contravened, with potentially significant consequences for researchers and participants. The relation between the ethical and the lawful is by no means exclusive to ethnographies directly committed to illegal practices, but working those edges assists us in extending broader questions about how we frame our research practice. This makes it all the more pressing to experiment with methods for dealing ethically with boundaries of the (il)legal. Only once the taboo surrounding illegal ethnographies is alleviated, can we, as individual researchers and as institutions, start living up to this task and the responsibility it unavoidably demands. To this end, there is much to be gained from looking to 'sibling' disciplines such as Criminology, Sociology and Anthropology which have long dealt with these issues, often with very high stakes, where it has long been understood that ethnography and transcendental ethical frameworks are incompatible. Recognising this on a more-than-speculative level, creating a situated ethics that will act as a bulwark to protect project participants, researchers and our institutions has never been more important.

Putting situated ethics into practice is, admittedly, not always an easy task, particularly for universities, which as legal entities may feel daunted by the threat of high legal costs and damage to public image 'illegal ethnographies' might present. As Elliott and Fleetwood (2017) show, for these reasons, ethnographers that face legal prosecution or police investigation across disciplines are still likely to receive little, if any, support from the universities that employ them. And yet, as we have illustrated in this chapter, the stakes are high if universities seek to maintain an open research culture dedicated to the value of knowledge, however uncomfortable.

In conclusion, we highlight five steps universities could take. First, rather than ignoring the possible legal dangers of ethnographic research, universities could provide teaching resources for staff and implementing relevant 'ethical dilemmas' into the ethics approval process. Second, universities could make legal support available to researchers conducting ethnographic practices into (possibly) unlawful activity. This would involve providing legal training workshops to groups of staff or to give access to individualised legal counsel helping educate researchers on their and their research-participants' rights in the field and during arrests. We can learn from the field of journalism, where unions and news platforms provide either in-house legal support or offer free legal support through connections with a law firm. Third, like journalists' unions, universities could lobby governments to increase protections for researchers to continue doing work that might be legally complicated but have immeasurable social and policy value. Fourth, again taking inspiration from journalism where this is common practice, universities have an important role to play in not only promoting but helping to standardise the implementation of security protocols regarding data retention and communication with research participants.[14] Fifth, and perhaps most importantly, universities should not shy away from publicly defending researchers in court and through public media channels. In this light, the University of Minnesota Human Research Protection Program set the precedent for practicing situated ethics by openly defending the actions of Scott DeMuth.

The equation of ethics and law can never be safely assumed. Only through ongoing evaluation can we, as researchers and as institutions, works towards the ongoing practice of ethical research, including, and perhaps especially, when it brushes up against the law.

References

Allen, C.M. (2005) *An industrial geography of cocaine*. New York: Routledge.

American Association of Geographers (2009) 'Statement of professional ethics', http://www.aag.org/cs/about_aag/governance/statement_of_professional_ethics (accessed 02 February 2017).

Arendt, H. (1972) *Crises of the republic: lying in politics, civil disobedience on violence, thoughts on politics, and revolution*. San Diego: Harvest Books.

Bashir, N. (2018) 'Doing research in peoples' homes: fieldwork, ethics and safety – on the practical challenges of researching and representing life on the margins', *Qualitative Research*, 18: 638–53.

Bernasco, W., Johnson, S.D. and Ruiter, S. (2015) 'Learning where to offend: effects of past on future burglary locations', *Applied Geography*, 60: 120–9.

Bissell, D. (2010) 'Placing affective relations: uncertain geographies of pain', in B. Anderson and P. Harrison (eds) *Taking-place: nonrepresentational theories and geographies*. Farnham: Ashgate, pp. 79–98.

Braidotti, R. (2006) 'Affirmation versus vulnerability: on contemporary ethical debates', *Symposium: Canadian Journal of Continental Philosophy*, 10: 235–54.

Bryant, L.R. (2011) 'The ethics of the event: Deleuze and ethics without Arche', in N. Jun and D.W. Smith (eds) *Deleuze and ethics*. Edinburgh: Edinburgh University Press, pp. 21–43.

Caitlin, C., Sultana, F. and Pain, R. (2007) 'Participatory ethics: politics, practices, institutions', *ACME: An International Journal for Critical Geographies*, 6: 304–18.

---

14 See for instance the News Corp Australia's defence of a journalist whose house was raided: https://www.nuj.org.uk/news/police-raids-on-australian-broadcaster-and-journalists/ Accessed 10 July 2019.

Calkin, S. (2019) 'Towards a political geography of abortion', *Political Geography*, 69: 22–9.

Cloke, P., Cooke, P., Cursons, J., Milbourne, P. and Widdowfield, R. (2000) 'Ethics, reflexivity and research: encounters with homeless people', *Ethics, Place and Environment*, 3: 133–54.

Dally, K. and Kozeny-Pelling, C. (2016) 'Research ethics at Oxford', http://www.admin.ox.ac.uk/media/global/wwwadminoxacuk/localsites/uashomepages/uasconference/presentations/presentationssept2016/W17_Research_ethics_at_Oxford.pdf (accessed 02 February 2017).

Dekeyser, T. (2015) 'Why artists installed 600 fake adverts at COP21', *The Conversation*, 7 December, http://theconversation.com/why-artists-installed-600-fake-adverts-at-cop21-51925 (accessed 18 April 2017).

Dekeyser, T. (2021) 'Dismantling the advertising city: subvertising and the urban commons to come', *Environment and Planning D: Society and Space*, 39: 309–27.

Denscombe, M. (2012) *Research proposals: a practical guide*. Milton Keynes: Open University Press.

Dickens, L. and Butcher, M. (2016) 'Going public? Re-thinking visibility, ethics and recognition through participatory research praxis', *Transactions of the Institute of British Geographers*, 14: 528–40.

Dyer, S. and Demeritt, D. (2009) 'Un-ethical review? Why it is wrong to apply the medical model of research governance to human geography', *Progress in Human Geography*, 33: 46–64.

Economic and Social Research Council (2015) 'ESRC research data policy', http://www.esrc.ac.uk/files/about-us/policies-and-standards/esrc-research-data-policy/ (accessed 24 April 2017).

Elliott, T. and Fleetwood, J. (2017) 'Law for ethnographers', *Methodological Innovations*, 10: 1–13.

England, K. (1994) 'Getting personal: reflexivity, positionality, and feminist research', *The Professional Geographer*, 46: 80–9.

Falconer, R. and Kingham, S. (2007) '"Driving people crazy": a geography of boy racers in Christchurch, New Zealand', *New Zealand Geographer*, 63: 181–91.

Ferrell, J. (1998) 'Criminological verstehen: inside the immediacy of crime', in M.S. Hamm and J. Ferrell (eds) *Ethnography at the edge: crime, deviance, and field research*. Boston: Northeastern University Press, pp. 20–42.

Ferrell, J. and Hamm, M.S. (1998) 'True confession: crime, deviance and field research', in M.S. Hamm and J. Ferrell (eds) *Ethnography at the edge: crime, deviance, and field research*. Boston: Northeastern University Press, pp. 2–19.

Fuller, D. (1999) 'Part of the action, or 'going native'? Learning to cope with the 'politics of integration'', *Area*, 31: 221–8.

Garrett, B. (2013) *Explore everything: place-hacking the city*. London: Verso.

Garrett, B. (2014) 'Place-hacker Bradley Garrett: research at the edge of the law', *Times Higher Education*, 5 June, https://www.timeshighereducation.com/features/place-hacker-bradleygarrett- research-at-the-edge-of-the-law/2013717 (accessed 28 October 2017).

Gilson, E.C. (2011) 'Responsive becoming: ethics between Deleuze and feminism', in N. Jun and D.W. Smith (eds) *Deleuze and ethics*. Edinburgh: Edinburgh University Press, pp. 63–88.

Gregson, N. and Crang, M. (2017) 'Illicit economies: customary illegality, moral economies and circulation', *Transactions of the Institute of British Geographers*, 42: 206–19.

Hall, T. (2012) 'Geographies of the illicit: globalization and organized crime', *Progress in Human Geography*, 37: 366–85.

Hamm, M.S. and Ferrell, J. (eds) (1998) *Ethnography at the edge: crime, deviance, and field research*. Boston: Northeastern University Press.

Hasting, J.V. (2009) 'Geographies of state failure and sophistication in maritime piracy hijackings', *Political Geography*, 28: 213–23.

Horton, J. (2008) 'A 'sense of failure'? Everydayness and research ethics', *Children's Geographies*, 6: 363–83.

Hynes, M. (2013) 'The ethico-aesthetics of life: Guattari and the problem of bioethics', *Environment and Planning A*, 45: 1929–43.

Katz, C. (1994) 'Playing the field: questions of fieldwork in geography', *The Professional Geographer*, 46: 67–72.

Lee, C. (2013) *The energies of activism: rethinking agency in contemporary climate change activism*. PhD thesis Durham University, http://etheses.dur.ac.uk/6953/ (accessed 25 July 2017).

Madge, C. (2007) 'Developing a geographers' agenda for online research ethics', *Progress in Human Geography*, 31: 654–74.

Markey, S., Halseth, G. and Manson, D. (2010) 'Capacity, scale and place: pragmatic lessons for doing community-based research in the rural setting', *The Canadian Geographer*, 54: 158–76.

McAuliffe, C. and Iveson, K. (2011) 'Art and crime (and other things besides...): conceptualising graffiti in the city', *Geography Compass*, 5: 128–43.

Miller, M. and Tewksbury, R. (2010) 'The case for edge ethnography', *Journal of Criminal Justice Education*, 28: 488–502.

Mitchell, R.G. (2001) *Dancing at Armageddon: Survivalism and chaos in modern times*. Chicago: University of Chicago Press.

Nast, H. (1994) 'Opening remarks on 'Women in the Field'', *The Professional Geographer*, 46: 54–66.

Neyfakh, L. (2015) 'The ethics of ethnography', *Slate*, 18 June, http://www.slate.com/articles/news_and_politics/crime/2015/06/alice_goffman_s_on_the_run_is_the_sociologist_to_blame_for_the_inconsistencies.html (accessed 24 July 2017).

Popke, J. (2009) 'Geography and ethics: non-representational encounters, collective responsibility and economic difference', *Progress in Human Geography*, 33: 81–90.

Rose, G. (1997) 'Situating knowledges: positionality, reflexivities and other tactics', *Progress in Human Geography*, 21: 305–20.

Routledge, P. (2002) 'Travelling East as Walter Kurtz: identity, performance and collaboration in Goa, India', *Environment and Planning D: Society and Space*, 20: 477–98.

Routledge, P. (2009) 'Towards a relational ethics of struggle: embodiment, affinity, and affect', in R. Amster, A. DeLeon, L.A. Fernandez, A.J. Nocella II and D. Shannon (eds) *Contemporary anarchist studies: an introductory anthology of anarchy*. London: Routledge, pp. 82–92.

Smith, D.M. (2001) 'Geography and ethics: progress, or more of the same?', *Progress in Human Geography*, 25: 261–8.

Svalastog, A.-L. and Eriksson, S. (2010) 'You can use my name; you don't have to steal my story – a critique of anonymity in indigenous studies', *Developing World Bioethics*, 10: 104–10.

Thoreau, H.D. (2001) 'Civil disobedience', in R. Dillman (ed) *The major essays of Henry David Thoreau.* New York: Whitston Publishing, pp. 47–67.

Till, K.E. (2001) 'Returning home and to the field', *Geographical Review*, 91: 46–56.

Valentine, G. (2005) 'Geography and ethics: moral geographies? Ethical commitment in research and teaching', *Progress in Human Geography*, 29: 483–7.

Vasudevan, A. (2015) 'The makeshift city: towards a global geography of squatting', *Progress in Human Geography*, 39: 338–59.

Wainwright, M. (2009) 'Drax protestors plead climate change case to jury', *The Guardian*, 1 July, https://www.theguardian.com/environment/2009/jul/01/drax-protesters-climate-change-jury (accessed 28 October 2017).

Whatmore, S. (1997) 'Dissecting the autonomous self: hybrid cartographies for a relational ethics', *Environment and Planning D: Society and Space*, 15: 37–53.

White, C. and Bailey, C. (2004) 'Feminist knowledge and ethical concerns: towards a geography of situated ethics', *Espace Populations Societes*, 1: 131–41.

10 Researcher trauma

Considering the ethics, impacts and outcomes of research on researchers

Danielle Drozdzewski and Dale Dominey-Howes

Introduction

In human geography, as in many socially oriented disciplines, research ethics is part and parcel of the research process. While the educational institutions we work for often mandate that we apply for and follow research ethics guidelines, we are also schooled to think ethically about our research work – and rightly so. As Sultana (2007: 375) has noted, ethical concerns "permeate the entire process of the research, from conceptualization to dissemination ... researchers are especially mindful of negotiated ethics in the field". For example, common questions pertaining to ethics in the research process may include: how do we conduct our research; what are the possible implications of our research on our participants; how can we minimise impact to participants; and can research design influence research ethics? In Australia[1], from where we write, we have a comprehensive research ethics process. There is also a national statement on ethical conduct in human research, which covers four key concerns related to research merit and integrity, justice, beneficence and respect (NHRMC 2018). The statement "sets out values and principles that apply to all human research. It is essential that researchers and review bodies consider these values and principles and be satisfied that any research proposal addresses and reflects them" (NHMRC 2018). The national research ethics statement and our own understandings of research ethics, garnered through tertiary education in human geography and experience as researchers, coalesce, we argue, along a central thematic line – do no harm to participants. The importance of this theme is paramount. Those involved in our research have a right to be treated fairly and with dignity.

The centrality of this focus to discussions of research ethics has meant that little, if any attention, is paid to how research content and location may impact researchers. As we have stated in a special issue of *Emotion, Society*

---

1   When this chapter was first drafted, Danielle worked at the University of New South Wales, Sydney, Australia.

DOI: 10.4324/9780429507366-10

and Space on Researcher Trauma, "we all 'do' research, but in 'doing' research, we rarely spend time thinking about the outcome of that research on our own emotional wellbeing, let alone on our writing and analytical research practices" (Drozdzewski and Dominey-Howes 2015: 17), despite the fact that we are often trained in reflexive research practice. Augmenting the significance of a focus on how we cope through our research process is the fact that the scope of research topics and locations in human geography (as in many other disciplines) is broad and far reaching. It is because many human geographers research at the interface of conflict, rapid environmental change and natural disasters, human displacement and in borderlands, with socially underprivileged communities and at/in post-war and post-disaster landscapes (Fisher 2018; McGeachan 2018; Prouse 2018; Dominey-Howes 2018), that as a discipline we are very susceptible to encountering traumatic content and places, and dealing with "sensitive or difficult topics" (Dickson-Swift et al. 2007: 328; van Zijll de Jong et al. 2011).

We need a well-trained workforce equipped with the skills and capabilities to work with research content, people and places that may be traumatising for those conducting the research. In health, behavioural and medical sciences, including the clinical sciences, there is some level of "training" to help researchers anticipate, recognise, plan for and respond to researcher trauma (Pearlman and MacIan 1995; Baird and Kracen 2006). However, for the broader social sciences and humanities, including geography, there is an almost complete absence of recognition of researcher trauma, its outcomes and effects representing a significant gap in our understanding of the research process and its effects on the researcher workforce (Drozdzewski and Dominey-Howes 2015; Dominey-Howes and Drozdzewski 2016). Researcher trauma is impactful and deleterious to researchers. It is rarely mentioned during the research process and even more infrequently written into ethics review processes. Here, we proffer suggestions for why we think this lack of recognition exists and manifests in the neoliberal academy. We provide examples of how research ethics processes, and the academy more generally, could better account for researcher trauma. The chapter is structured as follows. First, we outline what we mean by researcher trauma drawing from a critical review of the existing literature on traumatic research, asking: "what is researcher trauma?". Second, we show how researcher trauma, research ethics, feminist praxis and the neoliberal university coalesce in research practice. Last, we outline a set of recommendations and trajectories for future research and professional practice.

What is researcher trauma?

Defining researcher trauma is a challenging task. In some ways, a definition is counterproductive because feeling trauma is subjective, context specific and personal (Măirean and Turliuc 2013). It may also depend on a researcher's position in the research project. For example, Klocker (2015: 39) noted that she was responsible for people who encountered traumatic content and who were in "direct contact with traumatised people". Though reading and translating traumatic narratives, she did come into direct contact with the trauma being discussed. Kiyimba and O'Reilly (2016) have argued that, through repetitive listening, researchers or professional transcriptionists may also experience trauma, distress and stress while transcribing, due to the sensitive issues raised in research transcripts. Other researchers note that the repetition and frequency of exposure to traumatic content and/or to places of trauma can augment the severity of their influence (Dominey-Howes 2015; Drozdzewski 2015; Eriksen 2017). Coles et al. (2014: 97) have also argued that the potential for impact of trauma-related research involves the "interplay between the individual's personality, personal history, social and cultural contexts, and the traumatic experience". Furthermore, the outcomes of exposure to trauma are multiple and "not easily categorized" (Drozdzewski and Dominey-Howes 2015: 18); they "represent part of a tapestry of experiences that do not necessarily correlate with hard clinical diagnoses". In resisting the provision of a "fixed" definition of researcher trauma here (and elsewhere), we anticipate and agitate for an ethos of care and understanding for the multitude of possible experiences and encounters human geographers may face in their research. These incidences should neither be devalued nor dismissed because they do not fit the confines of a more rigid and specific clinical definition. What we can is offer a range of interpretations – an assemblage of traumatic experiences (Drozdzewski and Dominey-Howes 2015), drawn from literature on vicarious or secondary trauma and its outcomes.

In professional, clinically oriented fields such as psychology and counselling, researcher trauma and the emotional burnout of professionals who spend their lives focused on the traumatic experiences of others, is well understood, documented and accounted for (for example, Pearlman and Mac Ian 1995; Coles et al. 2014). Writing from the perspective of the health sciences, Lerias and Byrne (2003: 130) have described vicarious trauma as "the response of those persons who have witnessed, been subject to explicit knowledge of or, had the responsibility to intervene in a seriously distressing or tragic event". Eriksen (2017: 274) warns that "the lower intensity at which vicarious traumatisation can occur (compared with direct trauma) means many do not realise they are being affected". For example, Lerias and Byrne's (2003: 136–137) note: "victims may still be able to function relatively well in their life while still suffering its symptoms … [they] are often overlooked because their level of distress may not be significant enough to come to the attention of clinicians". Figley (1995: xv) has outlined that people can be "traumatized without actually being physically harmed or threatened with harm", but can experience compassion fatigue, the symptoms of which are "identical to secondary stress disorder (STSD) and equivalent to post-traumatic stress disorder". From the perspective of counselling and psychological health services, Pearlman and Saakvitne (1995: 31) describe vicarious trauma as "the inner transformation that occurs in the inner experience of the therapist [or other professional] that comes about as a result of empathic engagement with clients' trauma material". The transformative role of traumatic content in terms of enhancing individuals' capacity for empathy is a key tool for psychologists and counsellors. Similarly, Dickson-Swift et al. (2009) shift the focus to outcomes for the researcher in attempts to describe why we may experience and react to traumatic research content. They have contended that "as qualitative researchers, our goal is to see the world through someone else's eyes, using ourselves as a research instrument; it thus follows that we must experience our research 'both intellectually and emotionally'" (Gilbert 2001: 9; Dickson-Swift et al. 2009: 62). In our own critical review of literature relating to researcher trauma, vicarious trauma and STSD, we noted a range of symptoms "that may include but [is] not limited to headaches, sleep disturbances, gastrointestinal upsets, increased stress, loss of appetite, anxiety, depression, weight loss/gain and in extreme cases, post-traumatic stress disorder (PTSD) and tragically even suicide. Clearly, 'researcher trauma' has negative effects on the short and long-term wellbeing of researchers and their sense of self" (Drozdzewski and Dominey-Howes 2015: 18; Eriksen 2017).

Tying threads: researcher trauma, research ethics, feminist praxis and the neoliberal university

These encounters with researcher trauma garnered from our positions as researchers, colleagues and supervisors, are discussed in the sections that follow. Tying the threads of researcher trauma, research ethics, feminist praxis and the neoliberal university together, has been an iterative process – itself dependent on our own ability to exercise reflexivity in our research careers, but to build these understandings discursively and in collaboration. We begin this section by reflecting on the intersectionality of our own research agendas across these thematics.

Feminist praxis

Danielle's interest in research ethics stems both from her personal recognition that her research content and places of research have had impact on how she approaches new work in these places and deals with existing content (Drozdzewski 2015). Her research focuses on the interlinkages between place, memory and identity and how these are operationalised and maintained in post-war and post-conflict landscapes. She was also until recently a panel member of a human research ethics committee at her former university. Both stimuli have made her more aware of the process and practice of research ethics *and* the paucity of recognition, at the policy level, of the impacts that research can have on the researcher. Since her doctoral work began, nearly twenty years ago, she has frequently encountered narratives of death, struggle, suffering and destruction. As part of her research, she has conducted field-based research to concentration camps, sites of murder and post-war landscapes, as well as engaging with this content via secondary data sources (Drozdzewski 2014; 2016). She has documented the impacts of these encounters (Drozdzewski 2015), paying attention to the fact that time is needed for reflexivity (Falconer Al-Hindi and Kawabata 2002) and for the impact of, particularly disturbing data to settle. Such strategies (reflexivity and time) have been integral to her capacity to continue working in this research area. Writing about her own researcher trauma has been challenging. It has required attention to "the intensities of emotions, affective cultural practices, bodily sensations and subtle shifts in meaning and practice", aspects that Moss and Donovan (2017: 228) have recognised in their call for writing intimacy into feminist geography.

As her career has progressed, Danielle has better understood how a feminist research praxis, focused on the "decentering of knowledge production" and the situating of knowledge production relates to how dominant power structures affect and have bearing on knowledge production (McDowell and Sharp 2016). As Haggerty (2004: 391) has cautioned, "the institutionalized production of knowledge never proceeds unencumbered". Thus, in approaching her research from a feminist standpoint, she has endeavoured to remain attentive to how she (re)produces knowledge within the university system and how this knowledge is encumbered both by the power of the system and by the multitude of positions she brings to the research. As Moss et al. (2002: 6) have noted, in feminist practice in human geography "thoughts about identity, subjectivity and self, require thinking through how to access salient information" to create (nuanced) understandings of "identity, subjectivity and self".

For Dale, whose career has focused on exploring the impacts and effects of disasters triggered by natural hazards on people and places, recognition of the impact of researcher trauma on him has come much later in his career (Dominey-Howes 2015). As we work and write together, he is very much being guided, mentored and trained by Danielle in thinking about and understanding the value of a feminist praxis for influencing his "doing" and thinking. It is a work in progress. While positionality and awareness of situated knowledge have long been part of feminist research in human geography, building these into ethical research practice has not always been straightforward. This absence is partly because we may not be aware of the potential for research content or places "close" to us to cause harm (Drozdzewski 2015). It may also not be apparent because the same protocols and procedures that compel us to consider the implications of our research to/for our participants are either lacking or completely absent from the formal paperwork we complete before we head into the field. A further implication of the paucity of discourse, policy and documentation on researcher trauma is that supervisors, colleagues, health and safety personnel at work and loved ones, may also not ask questions about the potential impacts of a planned project on the researcher. Because the idea of researcher trauma falls outside the remit of orthodox research processes (including the approval of research by an ethics panel), there lies scope for us here to "identify ways that the underlying epistemology biases and shapes research process and results" (Cope 2002: 44). Using a feminist approach to contextualise researcher trauma and its peripheral position in research ethics, means that we need to be attentive to power in the production of knowledge and "entails thinking about the context within which the research takes place" (Moss et al. 2002: 8). The context we both draw from is the Australian neoliberal university sector (Connell 2015; Enright et al. 2017). However, we imagine that the place-specific context of Australia adds only minimal layers of difference. It is the pace, quantification and metric-driven focus of contemporary tertiary education systems that, we think, influence why less attention is paid to the impact of research on the researcher. We do, however, know that the rigour of ethic review processes differs among universities globally, thus in the next section, we provide an overview of the system we work in.

Ethics and the speed of research practice in the neoliberal university

In Australia, university ethics applications are comprehensive and lengthy. The application forms require detailed information on the proposed project, including its aims and background, methodology, strategies for participant recruitment; statements relating the proposed use of data and its storage (to ensure participant anonymity); potential conflicts of interest (personal and from funding bodies); whether the research involves certain target groups such as pregnant women, persons under 18 years of age, Indigenous people, people with a disability; the proposed language of research; letter of support from participating organisations; and, formal Participant Information and Consent Forms (see for example, https://www.nhmrc.gov. au/research-policy/ethics/human-research-ethics-committees). The compilation of this material requires time and concerted clarity regarding the research proposed and how it will be conducted. Further, time is required to assess the applications by a research ethics panel. Panels are often discipline related and in the case of the University of New South Wales, all projects deemed "High Risk" (those that target specific groups presumed especially vulnerable) are assessed by a central panel. The process may take weeks, if not months. Each Australian university has an ethics review process (similar, if not the same as that described above) designed to uphold their institutional responsibility and adhere to the National Health and Medical Research Council's (NHMRC) 'National Statement on Ethical Conduct in Human Research', which states "each institution needs to be satisfied that: a) its human research meets relevant scholarly or scientific standards; b) those conducting its human research i) are either adequately experienced and qualified, or supervised, and ii) understand the need to assess risks to their own safety and that of participants" (NHMRC 2015: 68). As Eriksen (2017: 274) has identified, "risk assessments, first aid training, and teamwork all form a standard part of 'safe research practice'." Yet, rarely does interpretation of the NHMRC's statement on assessing risks to one's own safety encompass consideration of what a researcher may experience while actually doing the research, which as Eriksen goes on to suggest, may include exposure to "emotionally and politically charged narratives". While the ethics approval process at UNSW, where Danielle was a panel assessor, recently underwent significant streamlining and revision, it still does not specifically address the role of the researcher.

The same is the case at The University of Sydney, where Dale is employed, and the University of Wollongong, where Eriksen wrote from (see also Haggerty 2004 for a Canadian example and Cahill 2007 for detail on the American Institutional Review Board). Aligning with the NHRMC's National Statement, the only possible avenue for identifying risk to the researcher in the UNSW ethics form, for example, was in the section entitled "Research where there is a risk of harm". In that section, a researcher could select 'yes' to the prompt on psychological harms or social harms. While there is no specific prompting relating these questions to either the participant or researcher, the normative thinking here is certainly that these questions relate to research participants. Nonetheless, as Danielle has pointed out elsewhere (Drozdzewski 2015), even when the research ethics form specifically prompts researchers to consider themselves, a closeness to topic, situated knowledge and/or naivety, may influence how they answer these questions. It is our contention that institutional protocols pertaining to ethics review do not adequately consider the potential implications of the research on the researcher. This omission comes in spite of what Haggerty (2004: 391) has termed an "ethics creep" – the outward expansion of the regulatory ethics system and the intensification of regulation of research-related activities. We are mindful that our call for greater recognition of the researcher in ethics review processes should not be considered as part of this creep.

However, researchers are integral not just to the functioning of the university, but also to the production of new knowledge. Their health is paramount to the continued success of both and there are serious and negative effects of vicarious researcher trauma (Dickson-Swift et al. 2008; Dunn 1991; Eriksen 2017; McCann and Pearlman 1990; Mukherji et al. 2014). Coddington and Miceli-Voutsinas (2017: 55) caution that "trauma reminds us that there is literally no place "outside" of research – and conversely, no research that is "beyond" the body." Trauma embodies (researcher) bodies. Our argument is that, in not recognising the deleterious consequences to the research, universities are failing in their duty of care to researchers.

Enter the neoliberal university

While we have attended to discussing a feminist research praxis and the ethics review process itself, there is a further overarching component of this narrative that is crucial for understanding how and why researcher trauma is so often neglected, namely the neoliberal university as a site and context of knowledge production. Australian and international surveys of researcher workforces report that researchers are simultaneously facing wide-scale funding cuts and demands for higher productivity (the culture of publish or perish). This situation leads to burnout, stress, dysfunction, career dissatisfaction and lack of support for researchers (Court and Kinman 2008; Mark and Smith 2012; Newell and MacNeil 2010; Winefield et al. 2003; Smaglik et al. 2014; Klocker and Drozdzewski 2014). These difficulties are overrepresented in women and early career researchers (including doctoral students) attempting to establish their careers and credentials. The same university system that is driving Haggerty's (2004) ethics creep, is also the pilot of the neoliberal audit culture, one that demands more research quantum, larger research grants and an acceleration in the turnaround time between (field) research and publication (Mountz et al. 2015; Morrissey 2015). In this system, there is less time to enact the type of feminist research praxis that we have advocated above when encountering content and places that are disturbing and disrupting to the capacity of a researcher to keep working. Time is a key factor. Time is necessary for good research (Rea 2010), but it is also crucial to have time "to reflect on the research process, and then to come to terms with the concomitant outcomes of traumatic content and place" (Drozdzewski and Dominey-Howes 2015: 4). Time is necessary for self-care (Eriksen 2017). Time is also necessary for reflexivity. For both authors, retrospective reflexivity has enabled space between the content/places of research and a capacity to return to that content and those places (Bondi and Fewell 2017). Time to process has meant recognition that many of the issues we struggle with as researchers may be "hidden deep within our subconscious" (Rose 1997; Jackson 2005: 106). Reay (1996: 443), for example, argues that 'real' reflexivity involves "giving as full and honest an account of the research process as possible, explicating the position of the researcher in relation to the researched". Building on the theme of time, we flag the increasing noise generated by a small group of scholars agitating for slow scholarship (Eriksen 2017; Hartman and Darab 2012; Martell 2014; Mountz et al. 2015). Slow scholarship "represents both a commitment to good scholarship, teaching, and service and a collective feminist ethics of care that challenges the accelerated time and elitism of the neoliberal university" (Mountz et al. 2015: 1237). The idea of slow scholarship hinges at counterpoint with the audit culture of the neoliberal university. The demands of the neoliberal university are quantitative, finitely defined, categorised and assigned value through performance merits. Bergland (2018) has argued that the "overlooked acts and tasks, typically related to emotional labour, for example ... [should be] given more weight, importance and value in comparisons with traditionally highly valued types of work". She continues by stating that: "a feminist ethics of care, is based on the claim that care and support are just as important functions and outcomes in the university as are efficiency and more measurable outcomes" (Bergland 2018: 5).

Time, an ethics of care and support coalesce under the rubric of feminist research praxis, which not only involves understanding how this triad influences our participants but also ourselves as researchers. It extends the ethics review process beyond identifying structural risks back towards questions of moral, conscientious, principled – ethical – research practice. Negotiating such a practice is formidable. Mackinlay (2016: 201) has claimed that "[i]n the university, neo-liberalism perverts feminist ideals in the pursuit of profit", it places them as out of place. In conjuring a notion of being out of place, we return to a quote from Meghan Cope that we used earlier in this chapter "that the underlying epistemology biases and shapes research process and results" (Cope 2002: 44). The ethics review process is governed by the university. As Haggerty (2004: 410), drawing from Bauman (1993) has stated: "ethical relationships are characterized by an ongoing interrogation of the types of responsibilities that we might owe to others, and which cannot be reduced to a simple exercise in rule following, it becomes apparent that the application of many of the existing rules bears little relationship to ethical conduct whatsoever". The governing of the research process, including determinations of how ethics review processes take shape, does not routinely make space for (and in many cases, excludes) an assiduity for the types of feminist praxis that would enable researchers working with traumatic content and places, the time and access to resources that they would need to manage their encounters. In trying to expose how researcher trauma is related to the concept of feminist praxis, research ethics and the neoliberal university, we have reached a point in the chapter where we now take responsibility for suggesting a path forward. How can we write researcher trauma into research ethics, maintain a commitment to good scholarship and navigate successfully the corridors of the neoliberal university?

Writing researcher trauma into research ethics and an ethics of care

The ethics review process is a key avenue for pursuing greater recognition of researcher trauma. As the vast majority of researchers at higher education institutions are required to undertake ethics reviews, building one or more questions into existing applications is a robust first step. As we noted earlier, ethics applications are often concerned with identifying potential risks or harms, especially with specific questions that seek to pinpoint research practices that may impact participants. We would advocate that potential new questions about the researcher separated from questions about research participants be added. Having these questions stand alone would garner recognition – even if as an afterthought – about how the proposed project may have impact on the self. Ethics review processes would in this way be more closely linked to risk assessment. This tethering would address what Eriksen (2017: 276) has argued to be a "missing link in the university system: [...] the current disconnect between the [Human Research Ethics Committees] HREC and universities' respective Professional and Organisational Development Services and Workplace Health and Safety (WHS) units". She also contends, and we concur, that "one area where the scope of 'ethics creep' has the potential to benefit researchers is in an area of academia that needs more support, namely researcher mental health" (Eriksen 2017: 276). Her efforts at championing change in her prior institution show that change is possible though also layered with bureaucratic process.

While building an awareness of researcher trauma into ethics review is ideal, there are also other steps forward that challenge us to broaden our conception of ethical research practice to one that encompasses more than just completing pre-fieldwork paperwork. A trauma-informed research practice embodies elements of researcher self-care (Christopher and Maris 2010; Eriksen and Ditrich 2015; Eriksen 2017); a politics of care (Askins and Blazek 2017; Lawson 2007); research training for academics, students and managers (Dominey-Howes 2015); and, a research management toolkit (Dominey-Howes and Drozdzewski 2016). In line with our discussion of feminist research praxis, Lawson's (2007: 210) call for a 'feminist care ethics' reminds us of 'the absolute centrality of care to our human lives' and of the need to propagate caring research communities. In human geography, discourses of feminist care ethics, geographies of care and responsibility and the politics of care ethics are gaining currency (McEwan and Goodman 2010; Popke 2009), but less attention has focused on how these concepts apply to ourselves as researchers and to our own research communities. In discussing care ethics in the academy, Lawson (2007: 210) has shown how previous scholarly engagement 'link[s] the devaluation of caring work to restricted flourishing'. She argues that this work affirms the centrality of care to social life; including the normative imperative to centralise care rather than marginalise it. Amid the flurry of daily life in the academy and the persistent pressure of performance targets, publications and grants, how often do we take time to think about how our colleagues and students may be handling what we know are tough research trips or taxing research content? Failure to engage and understand researcher trauma and its triggers creates fragile and volatile research communities and potentially has flow-on effects to the types of research we conduct (Calgaro 2015; Drozdzewski and Dominey-Howes 2015; Court and Kinman 2008). Further, a possible unintended consequence might be that researchers subconsciously shy away from undertaking potentially valuable research on topics that they sense (but can't be sure), might cause them harm – thus robbing the academy of valuable new knowledge, insights and methods.

Concluding thoughts

Unintentionally, we arrive at a point where the onus of responsibility for supporting a wider recognition of researcher trauma is somewhat unclear. Is it the responsibility of national and institutional research ethics committees? Is it the responsibility of heads of school or the division of research? Is it our responsibility as researchers, colleagues and supervisors? Is it all of the above and others we have not listed? What we do know is that small steps, such as making space for a chapter on researcher trauma in a handbook of research ethics in human geography, are part of the process of change. We also know that in continuing to write about researcher trauma we are upholding a feminist research praxis and acknowledging the value of reflexivity, intimacy, subjectivity and the self as integral part of the research process.

Turning finally to you, our reader, it is highly likely that you have read this chapter because you feel it relates to some aspect of your own research and/or to your teaching about research practice. If that is the case, then we invite you to consider the issues we have raised for your own research and for that of others in your research communities. We encourage you to challenge your human ethic processes, policies and practices and to foster an institutional culture of self- and community-research care.

References

Al-Hindi, K.F. and Kawabata, H. (2002) 'Toward a more fully reflexive feminist geography', in P. Moss (ed) *Feminist Geography in Practice: Research and Methods*. Malden, MA: Wiley Blackwell, pp. 103–116.

Askins, K. and Blazek, M. (2017) 'Feeling our way: academia, emotions and a politics of care', *Social and Cultural Geography*, 18 (8): 1086–1105.

Baird, K. and Kracen, A. (2006) 'Vicarious traumatization and secondary traumatic stress: a research synthesis', *Counselling Psychology Quarterly*, 19 (2), 181–188.

Bauman, Z. (1993) *Postmodern Ethics*. Oxford: Blackman. Available from: http://www.kessler.co.uk/wp-content/uploads/2012/04/Bauman_Postmodern_Ethics.pdf.

Bergland, B. (2018) 'The incompatibility of neoliberal university structures and interdisciplinary knowledge: a feminist slow scholarship critique', *Educational Philosophy and Theory*, 50 (11): 1031–1036.

Bondi, L. and Fewell, J. (2017) 'Getting personal: a feminist argument for research aligned to therapeutic practice', *Counselling and Psychotherapy Research*, 17 (2): 113–122.

Cahill, C. (2007) 'Repositioning ethical commitments: participatory action research as a relational praxis of social change', *ACME: An International Journal for Critical Geographies*, 6 (3): 360–373.

Calgaro, E. (2015) 'If you are vulnerable and you know it raise your hand: experiences from working in post-tsunami Thailand', *Emotion, Space and Society*, 17: 45–54.

Christopher, J.C. and Maris, J.A. (2010) 'Integrating mindfulness as self-care into counselling and psychotherapy training', *Counselling and Psychotherapy Research*, 10 (2): 114–125.

Coddington, K. and Micieli-Voutsinas, J. (2017) 'On trauma, geography, and mobility: towards geographies of trauma', *Emotion, Space and Society*, 24: 52–56.

Coles, J., Astbury, J., Dartnall, E. and Limjerwala, S. (2014) 'A qualitative exploration of researcher trauma and researchers' responses to investigating sexual violence', *Violence Against Women*, 20 (1): 95–117.

Connell, R. (2015) 'Australian universities under neoliberal management: the deepening crisis', *International Higher Education*, 81: 23–25.

Cope, M. (2002) 'Feminist epistemology in geography', in P. Moss (ed) *Feminist Geography in Practice: Research and Methods*. Malden, MA: Wiley-Blackwell, pp. 43–56.

Court, S. and Kinman, G. (2008) *Tackling Stress in Higher Education*. London: The University and College Union.

Dickson-Swift, V., James, E.L., Kippen, S. and Liamputtong, P. (2007) 'Doing sensitive research: what challenges do qualitative researchers face?', *Qualitative Research*, 7 (3): 327–353.

Dickson-Swift, V., James, E.L., Kippen, S. and Liamputtong, P. (2008) 'Risk to researchers in qualitative research on sensitive topics: issues and strategies', *Qualitative Health Research*, 8 (1): 133–144.

Dickson-Swift, V., James, E.L., Kippen, S. and Liamputtong, P. (2009) 'Researching sensitive topics: qualitative research as emotion work', *Qualitative Research*, 9 (1): 61–79.

Dominey-Howes, D. (2015) 'Seeing "the dark passenger" – Reflections on the emotional trauma of conducting post-disaster research', *Emotion, Space and Society*, 17: 55–62.

Dominey-Howes, D. (2018) 'Hazards and disasters in the Anthropocene: some critical reflections for the future', *Geoscience Letters*, 5 (1): 7. https://doi.org/10.1186/s40562-018-0107-x.

Dominey-Howes, D. and Drozdzewski, D. (2016) 'Preparing researchers to manage traumatic research'. http://blogs.nature.com/naturejobs/2016/09/23/preparing-researchers-to-manage-traumatic-research/#more-10757 (accessed 23 November 2018).

Drozdzewski, D. (2014) 'When the everyday and the sacred collide: positioning Płaszów in the Kraków landscape', *Landscape Research*, 39 (3): 255–266.

Drozdzewski, D. (2015) 'Retrospective reflexivity: the residual and subliminal repercussions of researching war', *Emotion, Space and Society*, 17: 30–36.

Drozdzewski, D. (2016) 'Encountering memory in the everyday city', in D. Drozdzewski, S. De Nardi and E. Waterton (ed) *Memory, Place and Identity Commemoration and Remembrance of War and Conflict*. London: Routledge, pp. 27–57.

Drozdzewski, D. and Dominey-Howes, D. (2015) 'Research and trauma: understanding the impact of traumatic content and places on the researcher', *Emotion, Space and Society*, 17: 17–21.

Dunn, L. (1991) 'Research alert! qualitative research may be hazardous to your health!', *Qualitative Health Research*, 1 (3): 388–392.

Enright, E., Alfrey, L. and Rynne, S.B. (2017) 'Being and becoming an academic in the neoliberal university: a necessary conversation', *Sport, Education and Society*, 22 (1): 1–4.

Eriksen, C. (2017) 'Research ethics, trauma and self-care: reflections on disaster geographies', *Australian Geographer*, 48 (2): 273–278.

Eriksen, C. and Ditrich, T. (2015) 'The relevance of mindfulness practice for trauma-exposed disaster researchers', *Emotion, Space and Society*, 17: 63–69.

Figley, C.R. (ed) (1995) *Compassion fatigue: Coping with secondary traumatic stress disorder in those who treat the traumatized*. Philadelphia, PA: Brunner/Mazel.

Fisher, D.X.O. (2018) 'Situating border control: unpacking Spain's SIVE border surveillance assemblage', *Political Geography*, 65: 67–76.

Gilbert, K.R. (ed.) (2001) *The emotional nature of qualitative research*. London: CRC.

Haggerty, K.D. (2004) 'Ethics creep: governing social science research in the name of ethics', *Qualitative Sociology*, 27 (4): 391–414.

Hartman, Y. and Darab, S. (2012) 'A call for slow scholarship: a case study on the intensification of academic life and its implications for pedagogy', *Review of Education, Pedagogy, and Cultural Studies*, 34 (1–2): 49–60.

Jackson, P. (2005) 'Gender', in D. Atkinson, D. Sibley, and N. Washbourne (eds) *Cultural Geography: A Critical Dictionary of Key Concepts: A Critical Dictionary of Key Ideas*. New York: I.B. Tauris, pp. 103–108.

Kiyimba, N. and O'Reilly, M. (2016) 'The risk of secondary traumatic stress in the qualitative transcription process: a research not', *Qualitative Research*, 16 (4): 468–476.

Klocker, N. (2015) 'Participatory action research: the distress of (not) making a difference', *Emotion, Space and Society*, 17: 37–44.

Klocker, N. and Drozdzewski, D. (2012) 'Commentary: career progress relative to opportunity: how many papers is a baby "worth"?', *Environment and Planning A*, 44 (6): 1271–1277.

Lawson, V. (2007) 'Geographies of care and responsibility', *Annals of the Association of American Geographers*, 97 (1): 1–11.

Lerias, D. and Byrne, M.K. (2003) 'Vicarious traumatization: symptoms and predictors', *Stress and Health*, 19 (3): 129–138.

Mackinlay, E. (2016) *Teaching and Learning like a Feminist*. Rotterdam: Sense Publishers.

Măirean, C. and Turliuc, M.N. (2013) 'Predictors of vicarious trauma beliefs among medical staff', *Journal of Loss and Trauma*, 18 (5): 414–428.

Mark, G. and Smith, A.P. (2012) 'Effects of occupational stress, job characteristics, coping, and attributional style on the mental health and job satisfaction of university employees', *Anxiety, Stress, & Coping*, 25 (1): 63–78.

Martell, L. (2014) 'The slow university: inequality, power and alternatives', *Forum Qualitative Sozialforschung/Forum: Qualitative Social Research*, 15 (3). http://www.qualitative-research.net/index.php/fqs/article/view/2223/3692 (accessed 1 May 2019).

McCann, L. and Pearlman, L.A. (1990) 'Vicarious traumatization: a framework for understanding the psychological effects of working with victims', *Journal of Traumatic Stress*, 3 (1): 131–149.

McDowell, L. and Sharp, J. (ed) (2016) *Space, Gender, Knowledge: Feminist Readings: Feminist Readings*. London: Routledge.

McEwan, C. and Goodman, M.K. (2010) 'Place geography and the ethics of care: introductory remarks on the geographies of ethics, responsibility and care', *Ethics, Place & Environment*, 13 (2): 103–112.

McGeachan, C. (2018) '"A prison within a prison"? Examining the enfolding spatialities of care and control in the Barlinnie Special Unit', *Area*, 1–8. https://doi.org/10.1111/area.12447.

Morrissey, J. (2015) 'Regimes of performance: practices of the normalised self in the neoliberal university', *British Journal of Sociology of Education*, 36 (4): 614–634.

Moss, P., Al-Hindi, K. and Kawabata, H. (2002) *Feminist Geography in Practice: Research and Methods*. Cornwall: Wiley Blackwell.

Moss, P., and Donovan, C. (eds) (2016) Intimate research acts, in *Writing Intimacy into Feminist Geography*, Abingdon: Routledge, pp. 227–236.

Moss, P. and Prince, M. (2017) 'Helping traumatized warriors: mobilizing emotions, unsettling orders', *Emotion, Space and Society*, 24: 57–65.

Mountz, A., Bonds, A., Mansfield, B., Loyd, J., Hyndman, J., Walton-Roberts, M., Basu, R., Whitson, R., Hawkins, R., Hamilton, T. and Curran, W. (2015) 'For slow scholarship: a feminist politics of resistance through collective action in the neoliberal university', *ACME: An International Journal for Critical Geographies*, 14 (4): 1235–1259.

Mukherji, A., Ganapati, N. and Rahill, G. (2014) 'Expecting the unexpected: field research in post-disaster settings', *Natural Hazards*, 73: 805–828.

Newell, J.M. and MacNeil, G.A. (2010) 'Professional burnout, vicarious trauma, secondary traumatic stress, and compassion fatigue', *Best Practice in Mental Health*, 6 (2): 57–68.

NHRMC (2015) *National Statement on Ethical Conduct in Human Research*, accessed May 20, 2019, https://www.nhmrc.gov.au/about-us/publications/national-statement-ethical-conduct-human-research#block-views-block-file-attachments-content-block-1.

NHRMC (2018) *National Statement on Ethical Conduct in Human Research (2007) – Updated 2018, NHMRC*. Canberra: Commonwealth of Australia.

Pearlman, L.A. and Mac Ian, P. (1995) 'Vicarious traumatization: an empirical study of the effects of trauma work on trauma therapists', *Research and Practice*, 26 (6): 558–565.

Pearlman, L.A. and Saakvitne, K.W. (1995) *Trauma and the Therapist*. New York: Norton.

Popke, J. (2009) 'Geography and ethics: non-representational encounters, collective responsibility and economic difference', *Progress in Human Geography*, 33 (1): 81–90.

Prouse, C. (2018) 'Autoconstruction 2.0: social media contestations of racialized violence in Complexo do Alemão', *Antipode*, 50 (3): 621–640.

Rea, J. (2010) 'Enabling good ideas to become even better', *Advocate: Newsletter of the National Tertiary Education Union*, 17 (4): 2.

Reay, D. (1996) 'Dealing with difficult differences: reflexivity and social class in feminist research', *Feminism & Psychology*, 6 (3): 443–456.

Rose, G. (1997). 'Situating knowledges: positionality, reflexivities and other tactics', *Progress in Human Geography*, 21 (3): 305–320.

Smaglik, P., Kaplan, K., Kelly, S. and Penny, D. (2014) 'Job satisfaction: divided opinions', *Nature*, 513 (7517): 267–269.

Sultana, F. (2007) 'Reflexivity, positionality and participatory ethics: negotiating fieldwork dilemmas in international research', *ACME: An International Journal for Critical Geographies*, 6 (3): 374–385.

van Zijll de Jong, S., Dominey-Howes, D., Roman, C., Calgaro, E., Gero, A., Veland, S., Bird, D., Muliaina, T., Tuiloma-Sua, D. and Afioga, T. (2011) 'Process, practice and priorities – reflections on, and suggestions for, undertaking sensitive social reconnaissance research as part of an (UNESCO-IOC) International Tsunami Survey Team', *Earth Science Reviews*, 107: 174–192.

Winefield, A.H., Gillespie, N., Stough, C., Dua, J., Hapuarachchi, J. and Boyd, C. (2003) 'Occupational stress in Australian university staff: results from a national survey', *International Journal of Stress Management*, 10 (1): 51–63.

11 Practical ethics approaches for engaging ethical issues in research geography

Francis Harvey

Introduction

This chapter considers how to engage ethical issues of human geographical research pragmatically.[1] The emphasis lies in developing the skills to think about ethical issues arising in conjunction with information communication technologies. The approach presented here practiced for some years in teaching about GIS ethical issues and published with that focus several times (see DiBiase et al., 2012), accounts can directly account for technology-related changes that accompany the ubiquitous use of ICT in science and society. Further, the practical ethics approach described in this chapter, engages institutional dimensions, their transdisciplinary aspects, legal issues and their ethical dimensions in human geographical research. This chapter aims to equip instructors and students with an approach, the seven-step method, that they can use and refine in their research and geographical pedagogy to prepare students and researchers to address better the myriad ethical challenges that arise in human research geography.

The following section reviews the concepts of practical ethics and considers them distinctly different from philosophical ethics, which dominate philosophical and ethical engagements in the sciences, including geography. Following up the pragmatic bent of this approach, the section after that describes the seven-step method and offers examples. The first example is from published work using this approach in teaching GIS-Professional ethics. The second example describes how this approach can be used in hypothetical classroom situations to consider issues in using volunteered geographical information in a diary-based study of the geographical dimensions of young peoples' social network usage and assessing the ethical dimensions of data collection using a survey. The conclusion summarises critical aspects of the practical ethics seven-step approach and relates critical aspects of using this approach in the classroom.

1   This paper is substantially based on work conducted with David DiBiase and Dawn Wright in the NSF Project Ethics Education for Geospatial Professionals project (GEO-0734903).

DOI: 10.4324/9780429507366-11

Practical ethics

The breadth and depth of ethical issues, concerns and problems in the sciences has become breath-taking. A series of scandals in psychology research going back to the retraction of Marc Hauser's studies from 2002 (see summary in Harvard Magazine, 2012) marks a watershed that in the ensuing years led to the documentation of many ethical issues and far more retractions (Pimple 2014). While human geography has its share of engagements with ethical issues reaching back many years (Bunge 1973), this engagement has neither been sustained in its scholarship nor in its theoretical and methodological developments.

Taking up research ethics in professional education remains important for self-obvious reasons (Harvey 2012). For research scientists, the same reasons apply: to be taken seriously in scientific and societal developments, geographical researchers must hold themselves responsible for the quality and the accuracy of their work (Chrisman, 1984, Schuurman, 2004, Crampton 2010).

What these two terms, quality and accuracy, mean in the abstract has been and remains a theoretical issue of great relevance to philosophers considering how what we observe relates to what we find possible and what we make sense out of in the world (ontological and epistemological issues). While this important, it is also of relevance to prepare and refine our abilities to account in our research work and teaching for dimensions of our research that we make choices about, face the influences of our values, and for which we can be held responsible.

Practical ethics is a more recent framework largely developed in engineering programs in the United States to integrate an application-orientated approach into the higher education training of engineers. Due to National Science Foundation (NSF) support, it has been highly influential in guiding curricular discussion in several engineering fields and computer science. It has also been an important contributor to science, technology, engineering and mathematics curriculum discussions and professional education developments. The pragmatic emphasis is also the strongest distinction to theoretical ethics. Practical, also known as applied, ethicists consider the application of theoretical ethics in professions that seek to define themselves by the commitment to standards of ethical and often moral behaviour, among other factors (Quinn 2009).

The strengths of practical ethics in engagements with professional geographic issues (Harvey 2012) are relevant for pragmatic engagements with ethical issues of research geography. The emphasis on engaging topical issues and concrete research practices makes the results directly relevant for the development of sound ethical research and helps assure good communication about the research to broader scientific and public communities. The explicit empirical aspect of practical ethics is also of great relevance to the development of ethical foundations for further research and communication with peers about its philosophical underpinnings. The philosopher Kwame Appiah (Appiah 2008) writes in this regard:

> Morality is practical. In the end, it is about what to do and what to feel; how to respond to our own and the world's demands. And to apply norms, we must understand the empirical contexts in which we are applying them. No one denies that, in applying norms, you will need to know what, as an empirical matter, the effects of what you do will be on others.
>
> (Appiah 2008: 22).

This excerpt from Appiah's significant work on ethics in contemporary society highlights the strengths of practical ethical approaches that ground our research activities in the actual interventions and practices of research. Rather than find ethical theories or principles that seem to explain choices, applied ethics focuses on consideration of both concepts together with actual situations and activities, including institutional and political dimensions. In other words, practical ethics must be grounded in the constitutive activities of research. In the classroom setting, this is by its very nature removed from actual workplace situations to a degree that will depend on students' work experience and institutional arrangements. In a research group setting, the approaches of practical ethics are directly relevant to the research at hand. Considerations of the choices, values and responsibilities of involved actors can lead to alterations in research activities or plans to account for the ethical issues taken up by the group. The deliberation of the ethical issues considers multiple perspectives and the assumptions of various actors as well as ourselves in analysing a situation.

This strength of practical ethics is also relevant in the classroom, as it allows students to take up and requires an engagement with perspectives that otherwise may be marginalised. Depending on the overall pedagogy, these considerations can be enhanced and connected to the people involved, and marginalised, through excursions, role-playing exercises or guest speakers.

The seven-step approach for teaching

This section focuses on explaining how the seven-step approach developed by Davis (2000) and refined by DiBiase et al. (2012) can be used in teaching. These points are also relevant in using the seven-step approach in a research group for a project as it is quite conceivable that members of a research group will be learning about the project in a variety of ways that parallel the learning required of students for the successful use of this approach in teaching. The method recommended for integrating the practical ethics approach in a classroom is a case-study method. The 'case method' is a common pedagogical technique for strengthening the moral reasoning skills of students in business, medicine, law, engineering, and computer and information science (Davis 2000; Keefer and Ashley 2001; Quinn 2006, 2009). The case study method can use short cases or extremely long cases. The former holds an advantage of fitting well into the tight class lengths of modules and with limited instructor preparation time required with the disadvantage that participants will need to rely on experience and assumptions to complete the case. Longer cases, which are sometimes used as the central pedagogical instrument in some graduate programs, even over multiple semesters, avoid most of these downsides but can be quite complex to prepare. Previous work on professional ethics made an emphasis on short cases the best-suited to the pedagogical context of short instructional periods or workshops to engage the cases. Shorter cases function more as frameworks for a more abstract discussion of ethical issues, which of course is still highly relevant.

The approach developed in that project closely follows Davis' (2000) "seven-step guide to ethical decision making" (outlined below). Similar ethical analysis models have been suggested by Keefer and Ashley (2001) and others. This approach, based on a framework of issues and scenarios, is also suitable in continuing educational settings. The description in this section has been previously published in DiBiase et al. (2012) and elsewhere. Davis (2000) describes the seven-step discovery and deliberation approach as follows.

Step 1. State problem. For example, "there's something about this decision that makes me uncomfortable" or "do I have a conflict of interest?"

Step 2. Check facts. Many problems disappear upon closer examination of situation, while others change radically.

Step 3: Identify relevant factors. For example, persons involved, laws, professional code, other practical constraints.

Step 4: Develop a list of options. Be imaginative, try to avoid 'dilemma'; not 'yes' or 'no' but whom to go to, what to say.

Step 5: Test options. Use such tests as the following: Harm test: does this option do less harm than alternatives? Publicity test: would I want my choice of this option published in the newspaper? Defensibility test: could I defend choice of option before Congressional committee or committee of peers? Reversibility test: would I still think choice of this option good if I were adversely affected by it? Colleague test: what do my colleagues say when I describe my problem and suggest this option as my solution? Professional test: what might my profession's governing body or ethics committee say about this option? Organisation test: what does the company's ethics officer or legal counsel say about this?

Step 6: Make a choice based on steps 1–5.

Step 7: Review steps 1–6. What could you do to make it less likely that you would have to make such a decision again? Are there any precautions can you take as individual (announce your policy on question, change job, etc.)? Is there any way to have more support next time? Is there any way to change the organisation (for example, suggest policy change at next departmental meeting)?

Examples

The following example shows a sample case from an NSF-funded GIS professional Ethics project involving mapping and how they can be engaged with the seven-step approach (Dibiase et al. 2012). This case and others are available at the gisprofessionalethics.org website. The first case involves a mapping project focused on an ethnic minority in southern California. DiBiase et al. (2012) provide a more detailed presentation.

Case study: Mapping Muslim neighbourhoods

A GIS Professional employed as director of the Center for Risk and Economic Analysis of Terrorism Events (the Center) at the University of Southern California receives an inquiry from an officer of the Los Angeles Police Department (LAPD). The officer, Commander Michael P. Downing, seeks the laboratory's assistance in a 'community mapping' project whose purpose is to "lay out the geographic locations of the many different Muslim population groups around Los Angeles," and to "take a deeper look at their history, demographics, language, culture, ethnic breakdown, socio-economic status, and social interactions." The community mapping project is to be one component of a counter-terrorism initiative that aims to "identify communities, within the larger Muslim community, which may be susceptible to violent ideologically-based extremism ..." (Downing 2007: 7). The director invites Downing to send the laboratory a Request for Proposal (RFP).

Soon after the telephone contact, Commander Downing is invited to Washington DC to explain the LAPD plan to the US Senate Committee on Homeland Security and Governmental Affairs. Committee chairperson Senator Joseph Lieberman cites it, among other similar projects, as an example of effective local-level counter-terrorism strategy.

News of the Senate Hearing and the LAPD plan is reported by the major media outlets including the New York Times, KNBC Los Angeles, and National Public Radio. Within days, representatives of three local Muslim groups along with the American Civil Liberties Union sent a letter to Commander Downing expressing "grave concerns about efforts by the Los Angeles Police Department ('LAPD') to map Muslim communities in the Los Angeles area as part of its counter-terrorism program." The signatories argued that the community mapping project ... seems to be premised on the faulty notion that Muslims are more likely to commit violent acts than people of other faiths. Singling out individuals for investigation, surveillance, and data-gathering based on their religion constitutes religious profiling that is just as unlawful, ill-advised, and deeply offensive as racial profiling" (Natarajan 2007: 1).

Meanwhile, the LAPD's RFP arrives at the Center. The well-funded project will involve considerable GIS work and will provide support for both student interns and professional staff. However, the director worries about the unfavourable publicity and possible legal action that might attend the project given the allegations of racial profiling. How should the director respond to the RFP?

Analysis following the seven-step approach

The following presents the discovery and deliberation involved in using the seven-step approach to analyse these ethical issues.

Step 1: State problem

Analysis begins with a discovery of the ethical issues. There are a number of potential ethical issues with this case, among them: Would work provided in response to the LAPD's RFP align with the mission of the Center for Risk and Economic Analysis of Terrorism Events? Does the Center's mission conflict with the University's? Should possible legal action or negative publicity influence the director's decision to respond to the RFP? Will responding to the RFP alienate the University, the public, and more specifically the Muslim community? Could this project be considered racial profiling?

Step 2: Check facts

Discovery continues with detailed considerations of facts.

- Fact: The Center's mission is "improve our Nation's security through the development of advanced models and tools for the evaluation of the risks, costs and consequences of terrorism and to guide economically viable investments in homeland security."
- Fact: University prides itself as "pluralistic, welcoming outstanding men and women of every race, creed and background" and "private, unfettered by political control, strongly committed to academic freedom."
- Fact: The LAPD has specifically invited the Center director, a certified GIS Professional, to submit a proposal in response to the RFP.
- Fact: The LAPD hopes to identify Muslim neighbourhoods within the city's Muslim community that may "be susceptible to violent ideologically-based extremism" (Downing 2007: 7).
- Fact: Representatives from three local Muslim groups and the ACLU object to the mapping project, claiming racial profiling (Natarajan 2007).
- Fact: LAPD portrays the mapping project as a "Community Engagement Plan" and specifically rejects charges of profiling.
- Fact: "Racial profiling," according to one definition, "occurs whenever police routinely use race as a factor that, along with an accumulation of other factors, causes an officer to react with suspicion and take action" (Cleary 2000).

Step 3: Identify relevant factors

This step moves into deliberations, as the analysis of situations begins in depth. Do the mission statements of the Center and the University conflict? An organisation concerned with terrorism assessment seems an awkward fit within a university that prides itself in pluralism and independence from government influence, notwithstanding its commitment to academic freedom. The missions may in fact conflict, though on the surface they appear to be simply driven by separate objectives. Is the project really profiling, and in what sense does mapping constitute profiling? Given the definitions cited above, the proposed project may at least be unethical, and at the worst, illegal and unconstitutional. The project would be particularly problematic if, unlike public Census surveys, it involves identification of individuals or small groups suspected of potential terrorist activities. Is even the appearance of profiling more damaging to the University and the Center than the benefit of receiving the funding? If the funding primarily provides a short-term gain for what becomes a longer-term conflict with the media, public, and student body, the rationale for the Center's involvement is questionable. How would RFP deliverables differ from publicly available Census data? Downing described some of the information the community mapping project was to collect, including data on Muslim population groups, detailing their "history, demographics, language, culture, ethnic breakdown, socio-economic status, and social interaction" (Downing 2007: 7). Some but not all of these data are publicly available from the US Census Bureau. It can be assumed that the community mapping project would go beyond simply aggregating existing public data sets.

Who might be able to offer guidance to the director for further direction? The University's legal counsel, ethics officer or conflicts of interest board, and the University's own IRB can provide guidance. Professional societies may offer additional suggestions.

Step 4: Develop a list of options

After deliberating contextual issues, this step creates possible actions.

> Option #1: Submit a proposal in response to the RFP.
> Option #2: Don't submit a proposal.
> Option #3: Request a modification to the existing RFP, or respond in a manner that removes any suggestion that profiling will occur.

Step 5: Test options

This continues to explore actions by following the systematic use of tests.

> Option #1: Submit a proposal in response to the RFP.
> Harm test: If awarded, the results of the project might alienate a community, cause irreparable harm to the University, and contribute to profiling.

Publicity test: Negative publicity is likely, though clients and sponsors will approve.

Defensibility test: The Center can defend applying from within its own mission statement, but may struggle to do so within the confines of the University.

Reversibility test: If a member of the Muslim community, the director might have natural reservations about the focus and outcome of the project.

Colleague test: Center staff would likely support the decision. Other university colleagues might oppose it.

Professional test: There are numerous potential conflicts with the GISCI's Rules of Conduct.

Organisation test: The University's ethics officer/legal counsel may have serious reservations.

Option #2: Don't submit a proposal.

Harm test: Possible future support from sponsors jeopardised.

Publicity test: Negative publicity is unlikely, except among clients and sponsors.

Defensibility test: Somewhat hard to justify within the Center, easier within the university. Difficult to justify to the LAPD and DHS given mission.

Reversibility test: If the LAPD were to have disqualified the Center from bidding for the project, the university may have protested.

Colleague test: Center staff might view the decision as weak. Others within the university might as well, while others still would praise the decision.

Professional test: Avoids potential conflicts with the GISCI's Rules of Conduct.

Organisation test: Legal counsel and ethics officers would probably support this decision given the cited concerns or suggest that strict controls be placed on the work being performed.

Option #3: Request a modification to the existing RFP, or respond in a manner that removes any suggestion that profiling will occur.

Harm test: Causes LAPD to reconsider end-goals of RFP.

Publicity test: Public may not recognise the distinction or care.

Defensibility test: Defensible but calls the Center's existence somewhat into question.

Reversibility test: University's concerns would likely be alleviated if the RFP itself were modified.

Colleague test: Colleagues would likely support and respect such a decision.

Professional test: Avoids potential conflicts with the GISCI's Rules of Conduct.

Organisation test: This choice would probably be supported by legal counsel/ethics officer.

Step 6: Select choice based on steps 1–5

After the tests, the analysis leads to the selection of a choice.

> Option #3 – Request a modification to the existing RFP, or respond in a manner that removes any suggestion that profiling will occur.

Step 7: Review steps 1–6

It is possible that discovery and deliberation led to more than one option left in the running. Reviewing the steps can help fine-tune the analysis and develop a presentation of issues that is sound. The Center director should reflect on potential conflicts between the Center's mission and the University's. An ethics statement may help avoid future conflicts. The key issue is whether or not the project does indeed involve profiling. If I were the director, I would not submit a proposal unless I was confident that the project would not infringe on the rights and privacy to US citizens are entitled.

Case study: Release of data

This is another case involving discovery and deliberation of issues connected to mapping. The Federal Highway Administration of the US Department of Transportation maintains a national inventory of over 600,000 bridges. States are responsible for conducting periodic inspections of bridges and to report their condition to the FHWA. In response to requests from "non-governmental sources" the FHWA will disclose records from the bridge inventory, but not the locations of individual bridges (which are recorded as latitude and longitude coordinates). Following the collapse of a bridge the area that caused several fatalities and dozens of injuries, the consultant working with this data for an FHWA project receives a telephone inquiry from a reporter who wishes to map structurally deficient bridges in his state.

Analysis following the seven-step approach

Step 1: State problem

Stating the problem starts the process of discovery. FHWA data can be provided without locational information about individual bridges. A journalist is requesting the data with locational information.

Step 2: Check facts

Considering facts is part of deliberation.

> Fact: The FHWA prohibits the release of locational data from the bridge inventory.
> Fact: A bridge in the area collapsed, causing deaths and injuries.

Fact: The consultant has access to the data to complete the FHWA project. No other use is permitted.

Fact: The reporter wishes to make a map of structurally deficient bridges.

Step 3: Identify relevant factors

This step moves the analysis of situations as deliberations begins in depth. The key factor appears to be that the consultant has access to the bridge inventory data to complete a project alone. That is, the data cannot be released, based on the terms of the project agreement.

Step 4: Develop a list of options

In this step, deliberation moves to develop a list of possible actions.

Option #1: Provide the data to the reporter.

Option #2: Provide a map of bridges with structural problems to the reporter.

Option #3: Retain the data and refer the reporter to the FHWA.

Step 5: Test options

After deliberating contextual issues, this step defines possible actions.

Option #1: Provide the data to the reporter.

Harm test: Leads to FHWA cancelling consultant's contract and possible legal action.

Publicity test: Public would only get a map. How would the map show the data?

Defensibility test: Not defensible under the terms of the contract.

Reversibility test: Not good if I would be responsible for any of the bridges.

Colleague test: Colleagues may support principle, but question benefits.

Professional test: May lead to conflicts with professional organisation

Organisation test: This choice would not receive organisational support.

Option #2: Provide a map of bridges with structural problems to the reporter.

Harm test: Possible aggravation of the FHWA, even if permitted (which needs to be verified).

Publicity test: Public may disregard the map.

Defensibility test: Possibly acceptable, needs to be verified.

Reversibility test: Any concerns may be eliminated, if FHWA supports the publication of the map.

Colleague test: Colleagues may support approach, yet question the effort.

Professional test: Involves blurring boundaries between profession and newspaper.

Organisation test: This choice may receive organisational support, if contract allows for publication of results.

Step 6: Select choice based on steps 1–5

Option #3: Retain the data and refer the reporter to the FHWA.

Harm test: Assures FHWA contract and relationship to consultant are protected.

Publicity test: Not possible.

Defensibility test: Completely defensible under the terms of the contract.

Reversibility test: Reflects value of legal contracts.

Colleague test: Colleagues may support principle, but question lack of support for this issue.

Professional test: May lead to conflicts with professional organisations.

Organisation test: This choice would certainly receive organisational support.

Step 7: Review steps 1–6

The review revisits discovery and deliberation activities. The consultant should review the contract and contact FHWA to determine if the provision of a map would be permitted. Also, reporter should be referred to FHWA to clarify the use of maps and possible acquisition of the data in the future.

A second example: designing a research project using an automated digital diary

In the context of research geography education, this second example assumes a grant has been awarded to study the geographical dimensions of young adult's social media usage. The project has received funding to develop a digital diary that participants would use to record their social media usage and the locations and times of this usage. From this broad scope, the project team has been asked to consider several ethical issues that the proposal pointed out. In this example, the specific issue is the proposed use of an "opt-out" clause in the agreement that participants would receive.

The team members suggest several scenarios for considering the ethical dimensions of such a blanket "opt-out" clause, which in the proposal would only be an option during the first use of the digital diary application. Following the seven-step approach, the following points express the deliberative development of ethical perspectives.

Step 1: State problem

The problem of "opt-out" is twofold. First, it impairs or even blocks participants' abilities to selectively choose when or where data is automatically recorded. Second, only available once, it cannot be altered except by deletion of the application, which would destroy any data, which was only on the device. The related matter of data synchronisation is an issue beyond the scope of these considerations, but necessary to note.

Step 2: Check facts

As the description of the problem makes understandable, the primary issue from the researcher's perspective is that the "opt-out" approach makes it less complicated for participants and assures that best-possible accurate data is collected. From the user perspective, the lack of an ability to regulate recording of social network activities and locations during the use of the diary could lead to a refusal to participate or grave concerns about the potential uses of the collected data. The impact on subject participation could have extremely negative consequences for this and future research projects. Based on changing perceptions of privacy, a solution that reflects legal requirements and cultural attitudes is deemed necessary.

Step 3: Identify the relevant factors

The members of the research project and all participants are the primary actors to consider. Further consideration of legal requirements and cultural attitudes mandates consultations with corresponding experts in the institution and peers who have considered these issues.

Step 4: Develop a list of options

Many options are relevant. A preliminary list has just two options:

1 Retain the "opt-out" approach, but introduce selective disabling of the diary.
2 Use an "opt-in" approach, which users must agree to at every initialisation of the diary application.

Step 5: Test options

Both options are compared using the seven tests Kennedy describes. The long-form results need to abbreviate the presentation. The gist is that both approaches are acceptable legally and culturally, but the "opt-in" approach affords participants greater control, which is central to the long-term nature of this research topic for the research group.

Step 6: Make a choice based on steps 1–5

The group chooses option two from step 4.

Step 7: Review steps 1–6

In reviewing the process of coming to this decision, the group finds it of great importance to develop a publication that explains the issues and describes a guideline for assessing the use of "opt-out" or "opt-in" provisions in research involving recording activity data of human subjects. The presentation of this work should receive a prominent position in the presentation of the group's research.

A third example: ethical concerns distributing a questionnaire to assess student attitudes towards campus surveillance

In this example, also in the context of research geography education, a proposed master's thesis focuses on attitudes towards CCTV surveillance of campus areas and classrooms. A student has developed a questionnaire and revised based on advisor input as a quantitative research technique. There were many changes and not enough time to consider ethical issues, especially those connected to the distribution of the questionnaires. To help prepare the ethics' assessment and submit the instrument with methods for approval to the college board of research review, the student decides to use the seven-step method to consider ethical issues in the distribution of the questionnaire, which are meant to be held between classes in front of lecture halls with randomly selected students.

The design of the questionnaire and concept for analysis follows a reproducible methodology and emphasises the quantitative analysis of questionnaire responses, but heightened sensitivity about CCTV surveillance and control and misuse of surveillance data has led to some rethinking, especially about its distribution on campus. In discussing the different places and student attitudes, it becomes obvious that "randomness" is harder to maintain in practice than in theory. Thinking about the potential biases, e.g., students coming from an electrical engineering lecture probably have a different range of opinions regarding CCTV surveillance than students coming from a human geography lecture, has led to doubts if a naïve "random" approach to distribution of the questionnaire is well-suited.

Step 1: State problem

The problem of how the random distribution and collection of questionnaires is ethically sound has two dimensions to consider here. First, the questions can have explicit or implicit biases that stimulate only narrow range of answers. Second, recipients of the survey are not random. Where and when the questionnaire students receive the questionnaire are key influences related to potential biases. Also, how the students who receive the survey are selected individually is a related issue. Finally, the place students get the questionnaire could influence the responses. Answers in a parking garage at 9 PM would certainly different than outside the bookstore at noon. Ethical engagement with the actual questions would be yet another issue to consider, but outside the scope of this study.

Step 2: Check facts

The simple, even naïve, random approach to finding and getting student responses seems good from quantitative data concepts, but the reality is that it is far more complex and the problems that can arise are real. The detailed distribution strategy needs to be assessed for potential biases and the strategy for soliciting participation needs to be better understood and prepared to address ethical concerns about the significance of the proposed research.

Step 3: Identify the relevant factors

Recent issues related to surveillance on campus and in the media can impact associations and preconceptions. This is unavoidable but getting to sound and ethically robust results means accounting for these issues. Some locations considered for distribution of the questionnaire may need to be reconsidered and possibly dropped. It is important to consider the ethical dimensions of biases in responses based on place, or the characteristics of the students that are at a place at a certain time. The ethical dimensions of methodological choices regarding distribution and solicitation of participation involve detailed discussions about how to account for potential biases.

Step 4: Develop a list of options

A number of options are relevant to refining distribution, soliciting participation in the survey and accounting for potential biases among students. The list has 3 options:

1 Instead of randomly speaking to students about participation, only ask every third student if they would participate. Keep track of the number of requests. After 15 minutes, see if female and male students were asked about the same number of times. If there is a bias, only ask students of the other gender.
2 Conduct a pilot study at one of the locations considered. See if there are any problems or concerns. Revise the distribution approach accordingly.
3 Verify that all locations are not subject to bias among the students. If a place-based bias is unavoidable, describe how the analysis takes this into account.

Different people suggest that there are other options, but these three seem germane to everyone.

Step 5: Test options

These three options need to be compared using the tests that Kennedy describes. Obviously, the publicity test is relevant, but the harm test can be relevant too, should the questionnaire disturb students. Regarding each option, it is imperative to consider them in light of possible adverse effects. Could the option be defended in public and would colleagues agree to the arguments and revisions to the distribution of the questionnaires. Since the questionnaire with description of the methodology goes for a review before the student can conduct the research, the student will know first-hand if the options satisfy organisational expectations and requirements.

Using the seven tests Kennedy describes the assessment of the three options points to the need of more knowledge about distribution, solic-itation and potential biases in terms of the research questions. Only the broader consideration would provide the robust assessment necessary to address conclusively the ethical dimensions. The gist of the tests with the given data is that changes to both distribution and solicitation are necessary because of the grave limits of the "random" distribution approach.

Step 6: Make a choice based on steps 1–5

The group choose option one and four from step 4. The student cannot con-duct the pilot study due to constraints for turning in the thesis.

Step 7: Review steps 1–6

In reviewing the process of coming to this decision, the group finds it of great importance to develop an ethical section of the thesis. Possibly, if there is sufficient interest, the development of more thorough research methodol-ogy guidance should account for these ethical issues and suggest the use of the seven-step approach to consider ethical dimensions more thoroughly.

Summary and conclusion

The practical ethics approach offers a constructive method for practically engaging current ethical aspects of geographical research. There are numer-ous challenges to developing pedagogy and examples to teach ethical issues in the diverse subdomains of geography.

In this context, pragmatic issues come to the fore in the use of the seven-step approach. However, it bears mention that the underlying social scientific matters how our research provides insight into human behav-iour are also considered. As Alfred North Whitehead (1925) reflected, scientific approaches that emphasise methodology may get world-changing results and may offer the basis for a comfortable advancement of science, but such progress can be deceptive, for the material emphasis diverts attention from the assertion of relationships between material and mind. Considerations of ethical issues in research pedagogy can facilitate a deeper and more significant connection between research theory, methods and results.

In this sense, the practical ethics approach can be very helpful to identify critical issues that are of significance beyond the scope of a research project. How these issues are engaged ethically may be well in alignment with commonly held values for a single project, but should be open for revisiting them and considering their refinement.

Critical evaluation

When thinking through pragmatic approaches to considering ethical issues in the classroom, the most significant challenge is just that many students will not have actual work experience involving ethical issues and every hypothetical situation has something which is contrived. This disadvantage can didactically be turned into strength. Both the use of role-playing or unconstrained reflections in the context of group discussions help to broaden the engagement with ethical issues, their contexts and open ways to go onto deeper considerations. Of course, group dynamics is critical in these processes and a zealous or malfeasant participant can lead the discussions into the problems, one can always expect from such persons.

More conceptually, the didactics of the engagements structured through the seven-step method move considerations past theoretical and philosophical issues and engage practical possibilities and the choices, values and responsibilities. Practically, a list of three aspects to a problem can help invigorate the discussion and help the group find the common grounds while acknowledging the breadth and potential for more assessment. A more theoretical orientation remains possible.

The use of this approach in the classroom requires significant time, as the process meanders and it is necessary to assure the discussions have adequate time. The length of the examples is a crucial factor. In this contribution, based on experience with GIS Professional Ethics project, short examples, which can be adequately discussed in the period of a single class, are best suited. Their disadvantage is a lack of specificity that can make it hard for under-supported students with little actual experience find themselves lost in the assumption they require specified.

These concerns are neither unusual in a classroom nor workshop. Addressing them is important, and especially essential in developing practical ethical approaches, which should provide means for students to deepen their understanding of the importance of ethical issues and pragmatically engage ethical issues of human geographical research.

Acknowledgements

Thanks to David DiBiase and Dawn Wright for discussions in the context of NSF Project Ethics Education for Geospatial Professionals project (GEO-0734903) that laid the foundation for this paper and other work on practical ethics. Discussions with Mike Kennedy, Chuch Huff and Mike Keefer helped develop the choices, values, responsibilities triadic concept. Portions of this chapter were written and developed during that project. The first case example is taken from published work (DiBiase et al. 2012) for GIS professionals and first published in that project at the UCGIS Summer Assembly 2010.

References

Appiah, K. A. (2008): 'Experiments in Ethics', Cambridge, MA: Harvard University Press.

Bunge, W. J. (1973): 'Ethics and Logic in Geography', in Directions in Geography, edited by Richard, J. C., London: Methuen and Company Ltd.

Chrisman, N. R. (1984): 'The Role of Quality Information in the Long Term Functioning of a GIS', Cartographica 21 (2): 79–87.

Cleary, J. (2000): 'Racial Profiling Studies in Law Enforcement: Issues and Methodology', Minnesota House of Representatives, Research Department, Retrieved from http://www.house.leg.state.mn.us/hrd/pubs/raceprof.pdf (accessed 3 March 2009).

Crampton, J. (2010): 'Mapping: a Critical Introduction to Cartography and GIS Malden', MA and Oxford: Wiley-Blackwell.

Davis, M. (2000): 'Ethics and the University', New York: Routledge.

DiBiase, D., Harvey, F., Goranson, C. and Wright, D. (2012): 'The GIS Professional Ethics Project: Practical Ethics for GIS Professionals', in Teaching Geographic Information Science and Technology in Higher Education, edited by Unwin, D. J., Foote, K. E., Tate, N. J. and DiBiase, D., Chichester, UK: John Wiley and Sons, pp. 199–209.

Downing, Michael P. (2007): 'Statement before the Committee on Homeland Security and Governmental Affairs, United States Senate', Washington DC, October 30. Retrieved from http://hsgac.senate.gov/public/index.cfm?Fuseaction=Hearings.Detail&HearingID=483590e6-9f4e-4aa6-b595-8ca3791e4acb (accessed 12 June 2008).

Harvey, F. (ed) (2012): 'Are There Fundamental Principles in Geographic Information Science?', Seattle: CreateSpace/Amazon Kindle.

Hauser, M. (2012): 'Engaged in Research Misconduct', in Harvard Magazine, Retrieved from https://harvardmagazine.com/2012/09/hauser-research-misconduct-reported (accessed 3 January 2018).

Keefer, M. and Ashley, K. D. (2001): 'Case-Based Approaches to Professional Ethics: a Systematic Comparison of Students' and Ethicists' Moral Reasoning', Journal of Moral Education 30 (4) (4): 377–98.

Natarajan, R. (2007): 'Letter to Commander Downing, LAPD', Retrieved from http://www.npr.org/templates/story/story.php?storyId=16162012 (accessed 12 June 2008).

Pimple, K. (ed). (2014): 'Emerging Pervasive Information Communication Technology', Berlin: Springer Verlag.

Quinn, M. J. (2006): 'On Teaching Computer Ethics within a Computer Science Department', Science and Engineering Ethics 12(2): 335–43. Retrieved from http://link.springer.com/article/10.1007/s11948-006-0032-9.

Quinn, M. (2009): 'Ethics for the Information Age', New York City: Pearson Addison Wesley.

Schuurman, N. (2004): 'GIS a Short Introduction', New York: Blackwell.

Whitehead, A. N. (1925): 'Science and the Modern World', New York: The MacMillan Company.

12 Facing moral dilemmas as a method

Teaching ethical research principles to geography students in higher education

Jeannine Wintzer and Christoph Baumann

Introduction

Field diary entries of Fabienne Kaufmann, Master's student at the Geographical Institute of the University of Berne (Kaufmann 2018):

Berne, autumn 2014: I have decided to write a Master's thesis that will create land cover and land use maps for Kenya. This will allow me to analyze satellite data, and I also hope that the fascinating and adventurous-sounding field trip will make it easier for me to enter the labor market in a development cooperation.

Kenya, March 2015: The feeling of not being integrated and not being at all able to integrate shapes my daily life in Kenya. Exchanges with the local population are limited to mutual observation. The gap between 'us' and the 'others' is reproduced daily through our behavior and the possibilities open to us. We explore the area by car, take pictures of settlements, people and nature and map them with our GPS. We hire people to escort and protect us, live in guarded areas and exercise in a fenced-in golf club.

Kenya, May 2015: Even after three months, the neighborhood children call us 'Mzungu, Mzungu', a term that refers to our white skin. I am afraid that I will constantly reproduce this mutual distance between the 'others' and 'us' in my Master's thesis.

Bern, September 2015: I have doubt about the original goal of the Master's thesis. In the field, on several occasions I was made aware that I should not take pictures of private plots of land. Especially not if it could be used for governmental purposes. I have respected that request. But now, in my workplace at the University of Berne, I am working with satellite data showing exactly those plots of land. I am working with this data without the knowledge of the persons concerned. I ask myself: 'How is this data legitimate and how can I use it for scientific purposes?'

The field diary entries cited here illustrate four geographical practices: First, surveying the earth, second, determining and attributing properties to territories, third, assigning persons and communities to local, regional or national spaces and fourth, in addition to the material products of

DOI: 10.4324/9780429507366-12

surveying, attributing and locating, and as a consequence of geographical practices, (re)producing the power of passing on information from 'others'. These practices of *world ordering* have little in common with the everyday life of local people in Kenya. Without their full knowledge, consent and involvement, a network of gradations is laid, spaces are classified and seemingly objective statements on physical and cultural geographic phenomena of the earth's surface are collected. The results are maps of land cover and land use as well as recommendations for action, for example, in the case of soil erosion. This is based on the idea that the phenomena of the world can be objectively mapped. In doing so, the powerful contexts and consequences of geographical practices are rarely reflected, although researchers in the field are confronted by them.

In this chapter, we reflect on some of the implications for ethics teaching that arise from situations such as the one described in the field notes above. This chapter is motivated by the following question: How can ethical principles are communicated to geography students in higher education in a way that enables students to translate taught principles into scientific as well as daily practice? It further asks: How should university teaching be structured so that students can identify their own ethical principles? The importance of this question lies in the fact that awareness of one's personal attitudes helps to make the decisions one takes comprehensible.

Here, a further question arises, namely that of how to successfully activate these principles for scientific research contexts when learning and research processes are for the most part temporally disconnected in university practices. Thus, while ethical principles[1] are mentioned in lectures and their importance is emphasised, they are then either forgotten when conducting research or the procedures necessary for ethically reflected action are not internalised. Consequently, it is necessary to debate which methods can be used to develop ethical principles so that knowledge of what 'should or should not be done' can be translated into the routines of geographical practice. These questions aim to sensitise students to the relationship between researchers and their research subjects and to ensure that they are able to interact with their research subjects in ways that respect ethical responsibilities.

To address these questions, section one discusses the need for ethical research principles in geography. Here it becomes clear that through the crisis of representation, the need for reflection on problems, data collection and analysis is growing as much as reflection on the interests of all actors involved. Section two presents approaches to moral education and moral

---

1   The following ethical principles apply to all scientific disciplines: Information, clarification, honesty, confidentiality, consent, privacy and anonymity, autonomy and self-determination, non-injury and fairness, impartiality and neutrality (Resolution des 60. Deutschen Hochschulverbandes, https://www.hochschulverband.de/779.html#; Gute wissenschaftliche Praxis. Denkschrift 'Sicherung guter wissenschaftlicher Praxis' http://www.dfg.de/foerderung/grundlagen_rahmenbedingungen/gwp/).

didactics, and, following on from this, advocates ethical principles and moral action for discussing moral dilemmas as a promising method. Finally, in section three, we present a concept for teaching ethical principles as well as integrating them into processes of understanding and practice. Moral dilemmas are situations in which several forms of action are possible. Any decision to take action, however, leads to a violation of a moral imperative. Discussing moral dilemmas is an intensive, argumentative examination of one's own moral principles as well as those of others, and promotes the competence of moral judgement. This proposes a didactic concept of teaching ethical principles that moves beyond frontal teaching and aims to trigger reflection of scientific practices as well as students' ability to act.

(Geographical) science and representation

In collecting data in Kenya and evaluating it in Switzerland, the Master student whose fieldnotes are cited above experiences moral dilemmas that result from the conflicting demands of fulfilling the research tasks set by the research group, while at the same time adhering to morally correct actions. She recognises that research does not take place in a vacuum, but rather is embedded in social and spatial contexts and, thus, in social processes. This insight concerns not only the research context in Kenya but also that in Switzerland. There, students frequently write Master's theses on topics that have been set by supervisors. So data collection and evaluation is preceded by the research questions and project goals previously defined and by the requirements of the funding body.

The student also recognises that the persons responsible for the project did not reflect on the research project, for example by questioning the goals pursued by the sponsors. In addition, no attempt appears to have been made to involve the local population by asking about any needs they might have, to which the researchers could provide answers and practice recommendations. Moreover, the fact that collecting data for sponsors and contractors to develop political strategies might have consequences for the local population had not been considered. Overall, the formulation of the problem, and the collection and use of data hold scientific potential, but can also entail a loss of power for the people affected by the research and the knowledge gained from it. As a result of this scientific practice, the student is increasingly confronted with the question of what is right. This question relates less to the correct application of specialist and methodological knowledge acquired in geographical studies than to the extent to which their research practice is ethically justified and beneficial or detrimental to those researched and to the researcher.

Postcolonial approaches (cf. Blunt & McEwan 2003) point out how deeply acquisition, production and transfer of knowledge are anchored within Western (scientific) ideals. For example, the analysis of historical contexts of knowledge production shows how strongly the European need for resources and people from the so-called protected areas directs scientific interest. These findings are also important for geography. Its institutionalisation as a scientific discipline at universities in the 19th century was closely associated with the knowledge required for fulfilling the needs of colonialism and imperialism, ranging from geographic composition to knowledge of resources and people. Before geography was established as a university discipline, geographers already played a major role in the way the world was perceived and acted upon. Referring to Edward Said's work Orientalism (1978) and his criticism of the practices of European sciences of the 18th century, Derek Gregory thus understands geography as a travelling practice and as a constellation of knowledge and power, a point that is illustrated in Fabienne Kaufmann's introductory example.

Research projects are increasingly evaluated according to the principles of ethical responsibility. This is a consequence of an epistemological shift that took place across all science disciplines between the late 19th and the early 20th century. Over the course of this shift, the objectivity of scientific knowledge acquisition and the associated idea of a clear representation of the world through science were questioned. It is not a review of geographic practices in the 19th century alone that would lead to the establishment of ethical standards. Also, the destructive aspects of science became increasingly apparent in the 19th and 20th centuries. For example, the use of machinery during industrialisation had led to high production rates, but also to unemployment and overproduction (Bernal [1939]1986). The development of the atomic bomb as well as chemical weapons is further examples of the increasing dangers of mechanisation for individuals, as witnessed in numerous military and political conflicts since the First World War.

At the beginning of their studies, students are usually unaware of the historical contexts of knowledge production and application as well as the necessary of an ethical perspective in science. During their initial semesters, students are rarely involved in research processes. Their role is that of consuming research results, and they know little about the development and implementation of research projects, or the consequences that scientific knowledge can have for local populations. However, as they progress in their studies, students become increasingly involved in research processes, and have to collect and evaluate data themselves and generate scientific results. Due to the growing importance of projects financed by third-party funding bodies, they are also often involved in their supervisors' research as auxiliary assistants, or directly involved by preparing their Bachelor's and Master's thesis projects. Some students then also go on to becoming directly involved in the production of geographical knowledge as doctoral candidates conducting their own research projects. Finally, in addition to their role as consumers and producers of geographical knowledge, it is worth remembering that geographers play a major role in transferring geographical knowledge as teachers and practitioners. This latter group, in particular, uses geographical concepts to propose solution strategies in their areas of work. It is therefore important to also reflect on the impacts of ethics teaching on the knowledges and practices of professionals who took geography degrees.

It is still all too often the case that ethical issues are only taken into consideration in geography for research purposes when people (or animals) are directly involved. However, it is not only in the field of human geography that ethical issues play a major role. Research questions with a physical geography orientation, such as measuring erosion or potential natural hazards, as well as surveying the earth's surface by means of satellite support, are only seemingly independent of society. They too involve data collection in socio-spatial contexts, as the above example shows. This can give rise to competing interests of local, national or international actors. It should also be born in mind that knowledge about the physical-geographical as well as the socio-geographical world cannot be gained context-free. Data on the world must be collected by researchers using analogous and/or technical survey instruments, and evaluated in its own right, and then with a view to existing knowledge. This means that scientific findings are not the result of objective observations of the world, nor are they merely representations of scientific and cultural facts. They are consequences of societal needs, so that science, as de-mystifier of the modern world becomes, itself, demystified (cf. Felt et al. 1995:8).

Is ethics teachable? Approaches to ethics didactics

The question of how far ethics can be taught is as old as the philosophical discipline of ethics itself. As Menon asked in the Platonic dialogue of the same name about 2400 years ago: "Can you tell me, Socrates, whether virtue is teachable?" (quoted after Hallich 2013:19, 70a). Before Socrates gets involved in a discussion of this question, he points out that it is necessary to first clarify what this virtue is. Here, we want to respond to the societal call for clarification of these terms by providing a miniature definition of what can be understood under ethical principles. Philosophy distinguishes between ethics and morality. Based on Hübner's definition (2013:11ff), ethics deals with the prerequisites and evaluation of human action. Thus, ethics is the process of thinking about legitimate action. Morality is the consequence of these preconditions, visible in actions. This definition refers to the concept of morality as a system of values in the sense of morally good versus morally bad. Thus, morality can be understood as a system of values consisting of individual values such as tolerance or equal rights, which is oriented towards human action and claims to be fully valid (Hübner 2013:13ff). Ethics is defined as the act of thinking about morality. To understand acting from a moral point of view is to evaluate it according to a logic of good/ethical/correct versus bad/unethical/wrong. This can be expressed in dimensions such as emotive consent or rejection, rational justification of the assessment and prescriptive invitation to choose or refrain from choosing a particular alternative (Hall 1983:5). In line with this view of ethics and morality, goals for university teaching can be defined as follows:

sensitising students to emotive involvement in morally relevant topics and increasing their capacity for ethical argumentation. This should help them to think about their actions and act in accordance with their moral evaluation (Hall 1983:6).

Lawrence Kohlberg's concept of dilemma discussion achieves the implementation of this goal. This is suitable as an instrument for recognising one's personal values as a first step, and as a second creating the ability to conduct an ethical discussion with other students, and then finally the ability to make a moral judgement to guide one's actions. Lawrence Kohlberg's theory of the development of moral judgement was developed in the late 1950s and has since been refined and modified, accompanied by the results of long-term studies, and discussed in its didactic implications (Köck 2002:122ff; Kohlberg 2001; Kohlberg 1996). Kohlberg assumes that moral development manifests itself in the development of moral judgement. By this, he means the ability to make an ethically justifiable decision in morally relevant situations (Köck 2002:122).

Based on empirical studies, he examines moral development in a level and graduated model (Figure 12.1). The empirical basis of level and level

Level	What is right?	Social perspective
I. Preconventional level: heterogeneous orientation; **Outward appearance of rules**		
1. Heteronomous morality	Obey rules because violation is punishable	Egocentric, only own perspective exists
2. Individualism	Obey rules when it serves the interests of someone (especially me)	Individualistic, there are other perspectives, but mine counts above all
II. Conventional level: social orientation; **Internalization of social rules**		
3. Mutual expectation of interpersonal conformity	Fulfillment of expectations from related parties	Common expectations take precedence over individual perspectives
4. Orientation to the social system	Comply with laws and social obligations	Perspective of the social system adopted
III. Post-conventional level: autonomous orientation; **Autonomy of rules**		
5. Articles of partnership for social utility	There are different values, but absolute fundamental rights apply	Perspective of a rational individual who can integrate different perspectives
6. Universal ethical principles	Orientation of self-imposed, universal moral rules	Perspective of the moral standpoint from which the social order is to be derived

Figure 12.1 Graduated model of development (shortened according to Kohlberg 2001: 38f).

identification is the so-called Moral Judgement Interview (MJI). Participants are confronted with moral dilemmas that they have to evaluate. The analysis of the interviews does not focus on the decisions of the participants but on their reasons. In total, Kohlberg divides the participants' reasons into three levels and six ranks. These differ in that the same morally relevant situations are evaluated with different argumentation strategies. Not only do the forms of argumentation become more differentiated and abstract, but the justifications are also characterised by increased broadening of social perspective and a questioning of value systems.

With this, Kohlberg designed a concept that contrasted with already existing ideas about moral didactics, because none of these fully address the issues of teaching ethical principles to motivate moral actions. In classical approaches of valuation, students were taught about ethics and morality through direct teaching. Lecturers teach morally correct behaviour and give students a "backpack full of virtues" (Kohlberg 1981, 9f, 31f, 184f, zit. in Oser 2001:70), which contributes to the development of their character. This approach is problematic, however. As early as 1928, Hugh Hartsthorne and Mark May showed in an empirical study of 850 pupils that there is no correlation between a continuous process of value mediation and situational moral action. Information about values does not automatically lead to the ability to make moral assessments and act competently. In addition, classical value mediation attempts to convey the competence for moral assessment, whereby students are given values that they have to follow ex cathedra. The mediated values themselves are not presented for discussion. According to Oser (2001:72), the acquisition of moral discernment and the willingness to act requires an "autonomous subject capable of meaning interpretation". This autonomy and ability to judge oneself is not promoted in direct teaching. Implicitly, this approach implies that there is a canon of universally valid values that are to be conveyed in lessons. However, values differ between historical phases and societies as well as societal groups, especially in individualistic and pluralistic societies (Hradil 1990; Beck & Beck-Gernsheim 2002).

A contrast to this approach of moral education is formed by the rejection of the classical determination of value. This rejection is based on the demand that education should be value-free and not aim to indoctrinate. The idea here is of the teacher as a sober, neutral expert, with a subject area that must be taught exclusively in its technical dimensions and is thus beyond any evaluative perspective. The problem of this approach is that technical topics are always associated with evaluative regulations. This applies not only to the fields of social science but also concerns disciplines like spatial planning, or research on hydrology, climate or natural catastrophe as these are not only characterised by technical, organisational and methodological aspects, but also by normative objectives. Concealing ethical dimensions is thus tantamount to asserting a form of protection, intended to remove the burden of having to launch a value-based questioning of topics, concepts and models.

Another position is represented by the approach of 'value analysis' (Oser 2001:72). This approach explicitly emphasises the normative dimensions of an object. Fixed, correct moral values are not prescribed here. Lecturers do not act as mediators but encourage students to discuss a topic from various angles and ethical points of view. To this end, lecturers refer to concepts of normative ethics that are applied to a problem. Thus, they may ask students to reflect on the question whether research should primarily achieve the greatest happiness for all, even if individual people/groups are instrumentalised, or whether scientific progress justifies all means of gaining knowledge. The approach of value analysis is reflective, and is suitable for university teaching, but presupposes, presupposed, that lecturers have a robust knowledge of philosophical and ethical concepts.

According to Kohlberg, the development of moral discernment always takes place in this fundamental sequence, which does not mean that everyone follows the development through to a post-conventional level. Kohlberg's theory is highly debatable and is open to criticism. Thus, it is not only a descriptive model but also normative in that the levels represent a hierarchy of 'morals'. For example, Kantian ethics (level 6) is clearly above utilitarian ethics (level 4, in part 5), which in turn is positioned above contractual morality (level 3). In a teaching situation, these problems can be counteracted by changing the order, for example through clustering the levels without numbering them, as a way of replacing the linear table. However, it represents an empirically supported and frequently tested framework for moral didactics, whereby moral dilemmas, in particular, play a central role (Lind 2009; Ulrich-Riedhammer & Applis 2013 for geography lessons in schools). We would therefore propose to adopt an approach that is based on Kohlberg's theory but applies it more flexibly. In the next section, we propose a method of teaching that takes moral dilemmas as a starting point for structured discussions following Kohlberg's framework for moral didactics.

Facing moral dilemmas as a method

The aim of confronting students with moral dilemmas is to encourage them to explicitly articulate and justify implicit moral concepts and to become able to discuss them in relation to the opinions of other students through argumentation. For this purpose, we present moral dilemmas in the following as a method. Moral dilemmas are characterised by conflict between at least two generally accepted ethical principles, in which there is no solution without undesirable consequences, i.e. consequences that are perceived by the peer group as problematic. As a rule, the dilemmas must be designed in such a way that they affect the students and inspire sympathy for the problematic situations faced by the represented protagonists. In Box 12.1, we suggest a set of questions, based on Lind (2009), which help to determine the principles of moral dilemmas. Furthermore, the task of lecturers during the discussion consists of acting as moderators. They should not evaluate the

---

Box 12.1

Construction principles of a moral dilemma (modified according to Lind 2009)

- Is there a morally relevant predicament?
- Is there anyone who finds himself in a predicament and has to make a decision about it?
- Does the story trigger empathy?
- Is the story understandable?
- Is the story not too long?
- Is the story realistic?
- Does the story allow different choices?
- Does the story not suggest a particular decision?
- Does the story address a topic that is relevant in terms of research practice?

---

individual assessments and methods of argumentation, but rather provide the framework conditions for discussion and ensure that they are adhered to (in particular, by concentrating on arguments, rather than on 'personal matters').

With reference to the construction principles, we present the following three examples of moral dilemmas (Box 12.2–12.4). In formulating these, a geographical reference was as important as the simple recognition of a conflict.

The experience with this method refers to five discussion periods: **(1) Surprise arises** – Due to assumptions of similar socialisation, origin and age, the students will naturally assume as a matter of course that the procedure is quite clear. Only the reactions from the group make it clear that there is no agreement at all among the students. **(2) Questions arise** – In the course

---

Box 12.2

Preparation and evaluation of soil erosion and land use maps

Within the framework of a Master's thesis project (scientific cooperation between the Kenyan state and Switzerland), soil erosion and utilisation maps for a region in Kenya are to be drawn up. You will be informed that a policy strategy to prevent erosion should be drawn up using the maps produced. You can see the high social relevance of your Master's thesis and decide to spend three months on site. After a short period of time, you come to realise that the local population perceives your presence as dangerous. Your photographs and surveys may be used for political purposes to collect arguments for reducing livestock herds or legitimising resettlement programs. How do you proceed?

Box 12.3

Sustainable agriculture and food security

You are writing a Master's thesis on new agricultural methods to guarantee food security in northern Indonesia. Together with a research group of ten people, you conduct fieldwork for two months. You are pleased to observe that the research project has a long tradition, which means that there is close professional exchange, but also private contact with the local population. In the course of your stay, your research group is often invited by local families to dinner or to take part in private celebrations. You appreciate this and take advantage of such opportunities to gain insight into the cultural traditions and everyday life of the population. When the three 9-year-old girls living in the village are to be circumcised, you are invited to the celebration. Within the research group there are discussions about whether to participate. How would you decide?

---

of this period, the students ask each other why a certain decision would be taken. From this, it transpires that the students are not aware of the basic factors underlying their opinions and cannot give concrete answers. Their reaction is along the lines of "well, that's how you do it". **(3) Demands arise** – Such answers are not satisfactory and give rise to demands such as: What do you mean by that? Why do you see it that way? What makes you think that? **(4) Arguments arise** – In consequence to these demands, arguments have to be introduced. These are the key to a student's fundamental principles in decision-making and acting and they form the basis for discussion. They

---

Box 12.4

Consequences of asylum status for refugee families

The city in which you are studying is building a settlement to house 250 refugees. The settlement will accommodate Ethiopian families with children for the duration of their asylum application process. As a social geographer, you are interested in their living conditions and decide to work on this topic in your Master's thesis. In order to answer your research questions, you plan to conduct 25 interviews with adult refugees. You do not speak the official language of Ethiopia (Amharic) and are therefore dependent on translators. Working closely with the head of the asylum centre, ten people are prepared to conduct a one-hour interview with you and a translator. During the interview survey, however, you perceive a deep reluctance on the part of the interviewees. They only give brief answers to the questions asked, and they all provide positive feedback about their experiences in the home. Your data appears one-sided and not representative. How do you proceed?

---

allow an exchange of opinions on daily or scientific matters. **(5) Insights arise** Students recognise the factors that underpin their decisions. This process allows them to contemplate whether these foundational factors are accepted, adapted or revolutionary, and whether they can change or maintain them.

These stages illustrate how a discussion of this type can promote a student's ability to consider situations from a moral point of view, become aware of and reflect on implicit ethical principles and exercise them in ethical discourse, i.e. a decision-making process determined by ethical arguments. The method of moral dilemmas offers a discussion without moral requirements of the lecturers, as would be the case with classical valuation. In this way, the five stages open up new perspectives beyond mere rules and regulations. In the course of the discussion, students also recognise the fundamental challenges of field research. This concerns power relations between the production of knowledge and the consumption of knowledge, as shown in Box 12.2, as well as power relations between investigative groups and managers of the asylum centre, as shown in Box 12.3. In addition, the students can recognise that research not only produces knowledge about an object but, as Box 12.4 suggests, can also change the attitudes of researchers. Last but not least, the students can conclude that the collection of data on site and the evaluation of data are usually spatially and temporally decoupled, that they are not objectively but socially integrated and that the results arise in specific socio-spatial context.

Conclusion

The field diary entries show that students are not well prepared for the ethical and moral challenges in the context of fieldwork. Three problems arise from this situation: firstly, the students are overburdened, so that uncertainties arise in the performance of their scientific work. Secondly, fear occurs because both – the students and the local population – feel threatened. Therefore, thirdly, central demands of post-colonial work such as non-hierarchical and participatory research cannot be redeemed. For these reasons, the paper calls for an adequate discussion about ethics and morality within university education and provides an example of teaching ethics and morality by confronting students with dilemmas. Kohlberg's theory of the development of moral judgement and the idea to confront students with dilemmas are promising. They emerge in contrast to value-free as well as value-based concepts of moral education because these don't fully address the issues of teaching ethical principles to motivate moral actions. However, this is important in order to internalise moral principles and to be able to carry out self-confident actions.

This, the internalisation of moral principles and ethical actions is one of the central demands of the Dublin Descriptors. Members of the Joint Quality Initiative working group made an essential contribution to including quality assurance and qualifications frameworks in the Bologna Process. Working within the framework of the Bologna reform, the group formulated expectations of the abilities that must be achieved for certain university degrees at the end of a Bologna study cycle. The so-called Dublin Descriptors formulate the competences to be achieved at the various levels of higher education (BA, MA, PhD), and are intended to serve as orientation for higher education institutions in developing their curricula and corresponding learning objectives at all reference levels. In addition to acquiring the skills of specialist knowledge and to apply them to new contexts, the Dublin Descriptors for Bachelor's degrees require students to be able to interpret scientific data under consideration of ethical aspects. For Master's degrees, students should have the ability to classify knowledge, master complexity and be able to make judgements within the framework of incomplete or limited information and under consideration of ethical responsibilities.

The Dublin Descriptors formalise existing demands in science for ethical principles. Ethical reflection of one's action is therefore a fundamental learning target for European universities and needs to be integrated into university teaching. Of course, ultimately the method of facing and discussing a moral dilemma continues to involve the staging of particular research contexts. Like any artificial situation, this lacks the emotions and uncertainties encountered in the foreign field, as well as unplanned situations.

References

Beck, U. and Beck-Gernsheim, E. (2002) *Individualization*, London: Sage.

Bernal, J.D. ([1939]1986) *Die soziale Funktion der Wissenschaft*, Köln: Pahl-Rugenstein.

Blunt, A. and McEwan, C. (2003) *Postcolonial Geographies*, Oxford: The Athlone Press.

Beltz, pp. 13–34.

Felt, U., Nowotny, H. and Taschner, K. (1995) *Wissenschaftsforschung: Eine Einführung*, Frankfurt am Main: Campus.

Hall, R. (1983) 'The teaching of moral values in geography', *Journal of Geography in Higher Education*, 7 (1): 3–13.

Hallich, O. (2013) *Platons 'Menon'*, Darmstadt: WBG.

Hradil, S. (1990) 'Individualisierung, Pluralisierung, Polarisierung: Was ist von den Schichten und Klassen geblieben?', in R. Hettlage (ed) *Die Bundesrepublik. Eine historische Bilanz*, München: C.H. Beck, pp. 111–138.

Hübner, D. (2013) 'Theorie der Ethik', in M. Fuchs and Th. Heinemann (eds) *Forschungsethik*, Stuttgart: J.B. Metzler, pp. 1–11.

Kaufmann, F. (2018) 'Positionalität in Forschungsprozessen. Eine (selbst-)reflexive Beobachtung', in J. Wintzer (ed) *Sozialraum erforschen: Qualitative Methoden in der Geographie*, Heidelberg: Springer, pp. 37–47.

Köck, P. (2002) *Handbuch des Ethikunterrichts*, Donauwörth: Auer.

Kohlberg, L. (1981) *The Philosophy of Moral Development Moral Stages and the Idea of Justice*, New York: Harper & Row.

Kohlberg, L. (1996) *Die Psychologie der Moralentwicklung*, Frankfurt am Main: Suhrkamp.

Kohlberg, L. (2001) 'Moralstufen und Moralerwerb. Der kognitiv-entwicklungs theoretische Ansatz', in W. Edelstein, F. Oser and P. Schuster (eds) *Moralische Erziehung in der Schule*, Weinheim: Beltz, pp. 35–61.

Lind, G. (2009) *Moral ist lehrbar: Handbuch zur Theorie und Praxis moralischer und demokratischer Bildung*, second edition, München: Oldenbourg.

Oser, F. (2001) 'Acht Strategien der Moral- und Werterziehung' in W. Edelstein, F. Oser, and P. Schuster (eds) *Moralische Erziehung in der Schule: Entwicklungspsychologie und pädagogische Praxis*, Weinheim: Beltz, pp. 63–89.

Said, E. (1978) *Orientalism*, London: Routledge & Kegan.

Ulrich-Riedhammer, M. and Applis, S. (2013) 'Ethisches Argumentieren als Herausforderung', *Praxis der Geographie*, 3: 24–29.

Internet sources

Dublin Descriptors, http://www.ecahe.eu/w/index.php/Dublin_Descriptors (accessed 25 August 2020).

13 Doing geography in classrooms

The ethical dimension of teaching and learning

Mirka Dickel and Fabian Pettig

Introduction

Explorative learning has been intensively discussed as an educational approach for quite some time. In this regard, the work of the *Bundesassistentenkonferenz* (BAK[1]) from 1970 is considered a 'milestone' in the German debate (Tremp and Eugster 2013: 392). The BAK builds on the ideas of Schleiermacher (2019/1808) and Humboldt (1960/1809,1810) for how students should engage with scientific research in universities. Over 200 years ago, both proclaimed that universities need to depart from the principle of instruction in favour of principles that allow students to actively engage with science. In a nutshell: students should learn exploratively, through research (Huber 2009: 4). The BAK's ideas for the design of educational contexts never lost relevance. Therefore, Reimann (2011: 291) considers it a *classic* rather than a *hype*. Along with a steady flow of thoughts on how to conduct explorative learning in educational contexts since 1970, Huber (2019: 20) further observes that, especially during the last 20 years, the rise of the term 'explorative learning' has simultaneously led to its diffusion. In this chapter, we follow Huber's general idea of explorative learning (ibid.), according to which educational contexts should be constructed in such a way that they start with learners' questions rather than following predetermined paths, for example, via tasks to be completed. This educational philosophy of explorative learning constitutes the starting point of our reflections on ethical dimensions of teaching and learning in geography.

Although the aforesaid originates from a debate on higher education, we will shift our focus on how to implement explorative learning in geography to school education. Our approach is guided by the following understanding of geography and of teaching: *Doing Geography* is, to us, not only the matrix of our everyday life, it is the matrix of geography lessons as well. The

---

1 Between 1968 and 1972 the BAK was the elected representation of nonprofessorial teaching staff in higher education in Germany. This board developed much-noticed concepts of reformation for tertiary institutes, first of all a document on learning through research in higher education.

DOI: 10.4324/9780429507366-13

way we talk about a specific phenomenon and make it a subject matter also means *doing geography*. That means that every decision in class is an ethical decision. Therefore there is a need to reflect on the ethical dimension of teaching and learning in schools (as well as in higher education, as demonstrated by Wintzer and Baumann, this book).

Geography lessons which are based on the explorative learning concept aim to enable pupils to ask their own questions about the lessons' topics and to confront these questions (methodically) in their own way, to learn to answer those questions and, then, to share and reflect on the results with others. Explorative learning entails more than practising (classical) scientific procedures in the classroom. This *more* should be understood as an *explorative attitude*, which should be initiated in geography lessons in such a way that topics are approached in a questioning and reflecting manner. School learning then becomes an experiential research process in which affective phenomena, which are phenomena that arouse the pupils' attention, lead students to ask their own questions, the reflexive answers to which, in turn, lead to knowledge.[2] In this interpretation, research and learning through explorative learning are closely related. Geography lessons that contribute to the formation of such an explorative attitude among pupils place special demands on the teacher who accompanies the (learning) experiences that arise in the course of the exploratory knowledge acquisition in the classroom.

The following discussion focuses on the guidelines that geography lessons which follow an educational philosophy based on explorative learning can use. Specifically, we explore the ethical dimensions that prime geography lessons which aim to facilitate explorative learning. For this purpose, we describe geography lessons as a complex activity field involving multiple decisions. The ethical dimension of geography teaching clearly cannot be determined according to a technocratic approach to teaching which pretends that geography lessons consist of many individual decisions that together comprise the learning environment. We adopt a different angle and think of exploratory learning in geography lessons as a coherent process, in which pupils *and* teachers gain (teaching) experience. To us, a good geography lesson opens pupils and teachers for a responsive experience, both in terms of *learning* (the subject-specific understanding of the subject), and *teaching* (transferring the subject-specific understanding). In the following sections, we outline in more detail what we understand as the *learning* and *teaching* aspects of a good geography lesson that follows an explorative learning approach. We then explain the structural elements of such lessons.

---

2   In previous research articles we highlighted concrete examples of experiential learning in educational contexts, e.g. on a field trip with students to Sylt (Dickel and Schneider 2013) and during a school project on critical cartography (Pettig 2019).

Geography lessons – an activity field of multiple decisions

For a distanced observer, geography lessons present themselves as an activity field of multiple decisions. It seems as if a teacher makes many individual decisions, which together make up the lessons' structure. Two levels which require decisions can be differentiated here: first is the question of how a topic can be adequately explored in the classroom (at the subject level). The second level concerns the design of the learning environment. In other words: how can the concrete learning environment be meaningfully designed such that pupils can adequately understand the topic (at the mediation level). For example, a distanced observer may ask how a teacher decides to adequately address a topic in class (material judgement). In view of the manifold observational modes and paradigms of geography as a subject, it is always possible to comprehend the teaching situation at different scale levels (local, regional, global), at different points in time (past, present, future), in different subject-specific dimensions of physical geography and human geography, and in relation to different understandings of space. In addition to this, teaching on many geographical subject matters requires some form of value judgement, i.e. the evaluation of a circumstance in relation to values of social sustainability. What can be considered as, for example, sustainable or socially acceptable, is also often contested and many geographically relevant topics, such as migration, scarcity of resources and justice, are controversially discussed due to the heterogeneity of social actors (including pupils themselves). In the course of globalisation, lifestyles and ways of life have diversified, such that there is a range of perceptions, awareness and interpretations that transform values.

In addition to the subject matter, teaching is also always about the 'how' of teaching. A teacher faces the question of how to design a learning environment responsibly such that pupils can engage meaningfully with the subject matter. However, there is no agreement even on this aspect of teaching, as teacher training programmes, internship guidelines and national or state education policies present trainee teachers with a plethora of approaches to teaching practices.

Given this complexity, it remains unclear how teachers direct their decisions in class. In order to obtain greater clarity on this, it is necessary to leave the distanced observer position and engage in teaching itself. We follow Gadamer's phenomenological hermeneutics here to argue that good (teaching-) practice cannot be observed from a distance. In order to assess whether an action is good and correct in a particular situation, it is necessary to consider one's own experience. The specifics of good practice can, therefore, only be experienced in lively situations. Consequently, we link the special feature of good practice to the practitioner's experience and ask for the competencies of the person who has practical experience and who aligns himself independently with his own experience. In other words: teachers' understandings of ethics and of ethical practice develop through experience.

Good teaching depends on an understanding that becomes apparent in the interaction itself. This understanding always underpins our 'worldly existence' ('In-der-Welt-Sein'). The practice, in this context, does not refer to actions in the sense of a mere choice between action options, but to actions in the sense of an ability. A teacher is able to refer in his or her teaching to his or her 'abilities', which increase over time (Neuweg 2004). This process of training ethical ability in praxis is itself referred to as ethicity.

The ethical dimension I: the subject-specific understanding of the topic

From a hermeneutic perspective, ethics does not have to be added to subject-specific teaching but any subject-specific teaching already has an ethical dimension, as it constitutes an *understanding encounter*. In concrete terms, this means that the manner in which a taught subject is made accessible to those who strive to understand it is itself an ethical practice. Understanding and decision are connected, not separate. They emerge together. The interest in *what* an appropriate understanding of the subject matter is, therefore, shifts to the question of *how* appropriate subject-specific understanding is becoming meaningful.

What does this mean exactly? The horizon of possibilities of how to understand subject-specific matter has expanded significantly since the cultural turn of the 1980s. As a consequence, it has become impossible to capture the subject matter of geography in a definitive way. Previous certainties have been replaced by a plurality of theoretical perspectives and methodological approaches that can be applied to the topic. This implies that the subject matter to be taught does not exist self-evidently either, but that it can only become a topic from a chosen perspective, i.e. in relation to the manner in which we refer to it theoretically and/or practically. In contrast to the scientific subject of geography, which is differentiated and fragmented into subdisciplines and different scientific approaches, didactics needs to reflect on which perspectives make sense in relation to the matter in hand.

To design subject instructions in *a* subject logic, would mean to socialise pupils into *a* manner of thinking and, thus, to conduct subject instruction as ideology. Critical and emancipatory teaching orients itself instead towards *manifold* scientific rationalities as its horizon of possibility.

The question that arises from this is how a subject becomes meaningful in the classroom. In order to fathom this, it is necessary to shift perspective from subject logics to didactic logics. We now think of a subject-specific understanding not only from the perspective of a subject classification but also from the perspective of the people who strive to understand a matter in a classroom discussion. Mediating subject knowledge via teaching is, therefore, not primarily concerned with transferring subject knowledge, but with mediating subject understanding. From a hermeneutic perspective, subject and pupils are related to each other in the process of understanding and this relationship is constantly transformed in the course of the encounter. Subject and pupils change simultaneously in this (educational) process. Hermeneutically, the subject matter of the lesson is always open to further interpretations. The ethical dimension of geographical understanding is manifested in this open horizon of dialogical mediation.

The ethical dimension II: transferring subject-specific understanding

A teacher contemplating how to make ethically sound decisions is like the proverbial millipede, who has had no trouble moving until he starts thinking about how he actually coordinates the complicated movement of his limbs and at that very moment becomes entangled and falls over. The story of the millipede points out that something happening intuitively causes problems when trying to understand it rationally. In the moment of trying to reflect via verbalising, one distances oneself from the self-evident action that is inherent in one's own habitus. During a spontaneous action, we access knowledge that intuitively guides us. However, it is so fundamental that it is not noticeable. In a lively debate, our actions are guided by 'tacit', inexpressible or implicit knowledge (Polanyi 2009/1966), i.e. by an intuitive ability in the sense of 'knowing-how' (Ryle 1949). For geography teaching, this means that ethical action during geography lessons cannot be conceived merely as the result of well-considered reflection. Although teaching actions that arise exclusively from well-considered justification may occur in everyday teaching, they are not typical of teaching practice.

In most cases, teachers act intuitively (Neuweg 2004; Matt-Windel 2014). For example, in the introduction to a topic, pupils may be asked to comment on a caricature. The teacher intuitively knows how long he or she will allow the pupils to express themselves, when he or she will interject with a thought-provoking question or comment, how he or she will cross-refer the pupils' remarks to each other and when he or she will end the commentary on the caricature. These actions do not arise from pure reflected judgement. Instead, the teacher will deal with the situation immediately. Rational reflection is, therefore, hardly effective *before* the action. Reflection only occurs after the action, for example, in-classroom meetings, to justify decisions already taken. In order to rethink what it means to act well as a geography teacher, it is necessary to recognise that we, as human beings, have always experienced the world in a physical and concrete manner. We can speak of good practice if the source of our insight and practice is an intellectual and resonant encounter with the other, a thing or a person. The ethical dimension is, therefore, not something that needs to be added to the subject-specific knowledge mediation. In concrete teaching situations, the debate of subject matters is inherently an ethical practice, if the encounter is of resonant quality. 'Resonance' (Rosa 2016; Rosa and Endres 2016) is then the category that is the norm for ethical decisions.

Facets of experience-oriented geography lessons

What do the above considerations mean, then, for how geography lessons should be organised? How should they aim to enable, accompany and reflect on experiences and which ethical skills do teachers need?

The structure of the experience is of crucial importance for this. Waldenfels (2002) describes human experience as a fragile accomplishment. Each experience occurs along the two poles of pathos and response. While the *pathic* moment of each experience denotes an unplannable, unpredictable event, as it is prereflective and not intentional, the *responsive* moment of each experience illustrates the individual's opportunity (and ability) of relating to these events, giving them meaning, and engaging with them. If the *pathic* is embraced and a manner of dealing with it is found, the event becomes an experience. These two poles of experience offer a number of connection possibilities for organising lessons. For if subject-specific teaching in the sense of exploratory learning aims at facilitating and accompanying experiences, i.e. if it is aimed at the subject-specific understanding of the subject matter along the way of finding and answering one's own questions, then spaces of opportunity must be created in which this can happen. What does this mean in concrete terms for the design of geography lessons?

We would argue that the following process should be taken into account in lesson planning: first, a viable example, which raises questions, must be chosen for the lesson; second, the example in the conversation must be subject-specific and, thus, become the topic of the lesson; third, the encounter between the pupil and the topic must be tactful, i.e. resonant, structured; fourth, an opportunity for an exchange via one's own experiences must be created. In the following, we illustrate these four facets of experiential geography teaching in detail.

Selecting a viable example

The serious involvement of pupils in the matter itself – a genuine commitment on the part of the individual – is necessary. This requires an example that has a certain reach. The reach of the example must be such that it enables the pupils to engage with it. In a lesson that answers essential questions, which arise for the pupils from the subject matter itself, a resonance experience is created in which pupils answer to the issues raised by the matter at hand. In these lessons, the questions and answers are not obvious, but emerge in the course of the discussion between teacher, pupils and the subject matter. The reach of the concrete example should be such that the pupils take the challenges of subject-specific learning seriously, rather than following the usual and rational course of action. A well-chosen example is, thus, the pivotal point of a lesson in which moral questions can be asked at all. According to Böhme, moral questions can be understood as those questions "which the individual will take seriously" (Böhme 1997: 111). A pupil who asks serious questions acts morally. Morality in school and in life is, thus, characterised by how something is done, or by the seriousness with which the individual engages with the matter, how he or she questions it and himself or herself. Each question calls for an original and determined manner of thinking in order to understand the unknown.

Giving the topic a subject-specific perspective

Every confrontation with an exemplary topic of geography lessons, not only lessons on explicit evaluation and assessment situations, trains ethical ability, on condition that it is explored resonantly. Geographical education, understood as a change or transformation of self and world relations, takes place on the basis of a concrete example in a micrological exploration. For this purpose, geography as a subject provides various theoretical lenses and research instruments that order the subject matter into patterns. In these different patterns, the subject matter appears in specific manners, since each pattern has its own visual acuities and blind spots. Bypassing through various patterns, which are brought into play during the lesson, the horizon of the subject matter's meaning, i.e. the way of thinking, through which the subject matter takes shape, changes and at the same time the individual pupil's relation to the world changes. Geography as a subject aims to bridge the gap between natural and social geography and, thus, between different logics of gaining knowledge. What seems like an impossible challenge at first sight can, however, succeed from a hermeneutic-phenomenological perspective. The latter makes it possible to grasp coherence in modernist ways of coding shared between the natural and social sciences that generate overlapping meaning constructions and ways of seeing the world. Tensions, fragilities and contradictions in the codings can be located and incorporated (didactically). In a humanistic understanding, the subject matter, as well as the pupil, change in the course of understanding if learning is conceived of as a dialogical confrontation in a principally never-ending process. This multidimensional geographical background makes it possible to gain differentiated subject-theoretical, socio-theoretical and cultural-theoretical as well as time-diagnostic patterns for geography teaching. These provide an orienting framework for initiating, shaping and reflecting on technical, practical and emancipatory knowledge mediation processes (see Vielhaber 1999).

Arranging a tactful interaction between the pupils and the subject matter

Geographical education via a lively interaction with the subject matter and a sensible encounter between pupil and subject matter means, above all, designing geographical education from the horizon of one's own geographical experience and understanding. In the interaction with the subject matter, one simultaneously gains one's own insights about the matter and about oneself; in other words, the understanding of something and the understanding of oneself go hand in hand, but do not come from within, because the interaction between the pupil and the subject matter is always a moment of chance. As explained by Bollnow (1955: 15): "It is always something that is independent of me and my planning, something that surprises me in the interaction, something that is coincidental and unpredictable". This affirmative and unpredictable something is capable of raising questions that require an answer. To create situations of interaction, thus, also always means to facilitate asking questions and finding answers during a lesson and, simultaneously, to acknowledge that interaction can be made possible didactically, but: interaction is certainly never guaranteed to occur on the basis of a didactic setting. Only through a praxis of reflexivity does the interaction unfold its inherent transformative potential. Geographical education, thus, means the exercise of mental independence, it means an increasingly profound understanding of the (life)world as a reference to meaning and a horizon of understanding that is constantly being transformed. This is an ethical process in a hermeneutical sense.

Reflecting on the experience process

The last step is reflection, i.e. the distanced reference to one's own process of experience. This is important for achieving geographical education because insights gained must also be experienced in their basic contingency in order to possibly contribute to self-education. In this phase of a geography lesson, questions that trigger further thought can be posed, such as which (subject-specific) perspective was taken in the course of the research, how did this decision shape how the subject matter was investigated, which views were (consciously) hidden? In this manner, the retrospective view can show how the subject matter was geographically revealed to the pupil, what these questions mean for the subject matter and also for the pupils and their access to it. Geographical education in this sense does not refer to learning subject-specific geographical methods and perspectives on the world, but consists of becoming reflexively aware of one's own being in the world and finding out something about oneself and one's relation to the world.

Conclusion

A geography lesson based on principles of experiential learning that is aimed at facilitating and accompanying experiences follows several interrelated motives. It is example based, enables interaction between pupils, teacher and subject, focuses on finding questions and giving answers and implies reflecting on complex experiences. Geography lessons thus conducted respond to both ethical dimensions: the multiperspective understanding of the matter, on the one hand, and the mediation of subject-specific comprehension on the other.

In such lessons, geographical education is understood as a process, i.e. as a continuous formation of world relations, and not as the result of cognitive knowledge transfer. In performing this transformation, the pupils proceed to self-existence. A resonance space is opened up in the course of practising self-existence and in developing the ability to act in a class (Rosa 2016: 411). Pupils' fulfilment of life is realised via increasing emancipation. In this sense, resonance is the key category for good geography lessons. The question of good teaching is an ethical question per se.

References

Böhme, G. (1997) *Ethik im Kontext*. Frankfurt a.M.: Suhrkamp.

Bollnow, O.-F. (1955) 'Begegnung und Bildung', *Zeitschrift für Pädagogik*, 1 (1): 10–32.

Dickel, M. and Schneider, A. (2013) 'Über Spuren. Geographie im Dialog', *Zeitschrift für Didaktik der Gesellschaftswissenschaften*, 1: 80–98.

Huber, L. (2009) 'Warum Forschendes Lernen nötig und möglich ist', in L. Huber, J. Hellmer and F. Schneider (eds) *Forschendes Lernen im Studium. Aktuelle Konzepte und Erfahrungen*. Bielefeld: UVW, pp. 9–35.

Huber, L. (2019) '"Forschende Haltung" und Reflexion: Forschendes Lernen als Thema, Ziel und Praxis der Lehrerinnen- und Lehrerbildung', in M. Knörzer, L. Förster, U. Franz and A. Hartinger (eds) *Forschendes Lernen im Sachunterricht. Probleme und Perspektiven des Sachunterrichts*, Band 29. Bad Heilbrunn: Klinkhardt, pp. 19–35.

Humboldt, W.v. (1960/1809,1810) "Denkschrift über die äußere und innere Organisation der höheren wissenschaftlichen Anstalten in Berlin', in W. Weischedel, W. Müller-Lauter and M. Theunissen (eds) Idee & Wirklichkeit einer Universität. Berlin: De Gruyter, pp. 193–202.

Matt-Windel, S. (2014) *Ungewisses, Unsicheres und Unbestimmtes. Eine phänomenologische Studie zum Pädagogischen in Hinsicht auf LehrerInnenbildung*. Stuttgart: ibidem.

Neuweg, G. H. (2004) *Könnerschaft und implizites Wissen: zur lehr-lerntheoretischen Bedeutung der Erkenntnis-und Wissenstheorie Michael Polanyis*. Münster: Waxmann Verlag.

Pettig, F. (2019) *Kartographische Streifzüge. Ein Baustein zur phänomenologischen Grundlegunge der Geographiedidaktik*. Bielefeld: Transcript.

Polanyi, M. (2009/1966) *The Tacit Dimension*. Chicago: University Press.

Reimann, G. (2011) 'Forschendes Lernen und wissenschaftliches Prüfen: die potentielle und faktische Rolle der digitalen Medien', in T. Meyer, C. Schwalbe, W.-H. Tan and R. Appelt (eds) *Medien & Bildung*. Wiesbaden: Springer, pp. 291–306.

Rosa, H. (2016) *Resonanz. Eine Soziologie der Weltbeziehung*. Frankfurt a.M.: Suhrkamp.

Rosa, H. and Endres, W. (2016) *Resonanzpädagogik. Wenn es im Klassenzimmer knistert*. 2nd edition. Weinheim and Basel: Beltz.

Ryle, G. (1949) *The Concept of Mind*. New York, Melbourne, Sydney and Capetown: Hutchinson.

Schleiermacher, F. (2019/1808) *Gelegentliche Gedanken über Universitäten in deutschem Sinn*. Berlin: De Gruyter.

Tremp, P. and Eugster, B. (2013) 'Forschungsorientiertes Studium – Forschendes Lernen: Ausgewählte Literaturhinweise', *Beiträge zur Lehrerinnen- und Lehrerbildung*, 31 (3): 389–395.

Vielhaber, C. (1999) 'Vermittlung und Interesse – Zwei Schlüsselkategorien fachdidaktischer Grundlagen im Geographieunterricht', in C. Vielhaber (ed) *Fachdidaktik kreuz und quer. Materialien zur Didaktik der Geographie und Wirtschaftskunde*, Band 15. Wien: Institut für Geographie und Regionalforschung, pp. 9–26.

Waldenfels, B. (2002) *Bruchlinien der Erfahrung*. Frankfurt a.M.: Suhrkamp.

14 Ethics of reflection

A directional perspective

Matthew G. Hannah

Introduction

The core argument of this chapter is that where we direct our critical, analytical attention and engagements, and where we do not, already fundamentally shapes ethical aspects of our work, quite apart from substantive ethical issues relating to its specific content. We are unable to avoid reifying, essentialising, objectifying a wide range of potential matters of concern simply because we are turned towards others at the moment. All possible epistemological positions are circumscribed in a practical sense by our inability to turn our attention critically towards all things at once. Most academics are at least latently aware that the selective directedness of what we do is important. However, attempts to think through the implications have been rare. This chapter seeks to address the selective directedness of academic practice and to illustrate its implications by reflecting on the practice of reflection as a basis for ethically defensible academic work. Doing so leads in turn to a directional perspective upon the unobtrusive but very important practice of reviewing the work of others. Reviewing is a genre of academic writing in which our expectations about responsible and conscientious academic work – both by others and, implicitly, by ourselves – are explicitly expressed. Thus it offers a good opportunity to illustrate the effects of ignoring the directional selectivity of everything we do. The kind of argument made with respect to reviewing could be made in many other areas of academic work as well.

Reflection serves two main functions in social scientific research, one epistemological, the other ethical. At the first, epistemological level, at least the following three kinds of reflection are important: (1) reflection upon how the conception, design and types of results of a research project relate to the project's wider social, economic, political and cultural contexts as well as to the positionality and interests of the researcher; (2) reflection upon the epistemological limitations and the social effects of data-gathering practices and forms of recording in the field and (3) reflection upon the linguistic and cultural lenses, including statistical and other seemingly neutral or technical forms of representation, through which empirical material is interpreted

DOI: 10.4324/9780429507366-14

and the results represented. To the extent that these forms of reflection are pursued carefully and documented systematically, they serve the same function as do the more formal procedures of experimental documentation and hypothetical-deductive method in the natural sciences: making research as transparent as possible.

Reflection is also understood to be ethically relevant because it helps researchers avoid treating and representing the human subjects of research in ways that objectify or reduce the independent complexity and integrity of their lives and social relations. 'Ethical' aspects of research in this sense are also inherently 'political' because what reflective practices aim to avoid is the establishment or reproduction of unnecessarily hierarchical power relations between researcher and research subjects. This chapter is concerned primarily with the ethical (and thus also political) implications of reflection. It should be kept in mind, however, that especially in what can loosely be termed 'critical social science', which includes much research carried out at the cutting edge of human geography today, the ethical and the epistemological are closely intertwined. The first section of this chapter surveys different approaches to reflection as an ethical strategy, discussing the ideas of Pierre Bourdieu, Donna Haraway, Gillian Rose and Karen Barad. The first two authors recommend some variant of reflection as a means of exercising responsibility in research; the latter two authors offer constructive critiques of it. All four treat reflection chiefly as a matter of separation or distancing. From a directional perspective, it becomes clear that understanding reflection mainly as a question of distancing yields an incomplete picture. Reflection involves both separation or distancing and *turning*. Reflection is one of many directed academic practices, and when we engage in these practices, we turn towards some matters while turning away from other matters. To bring out the turning aspect of reflection more starkly, the second section discusses recent work on the economic and political significance of attention. This work makes it clear that the question of where we direct our finite attentional capacities involves an ethics different from but underlying ethical aspects of the substance of what we think, say, write or do. Put very broadly, it is an ethics of where, not what. The third section returns to the accounts of reflection previously summarised. Its aim is to show that reflection as traditionally understood, as well as attempts to move beyond reflection, both necessarily rely on a reflective moment of turning back. This is brought out with reference to Bourdieu and Barad. The fourth and final section applies a heightened sensitivity to the element of turning in reflection to practices and criteria of reviewing and evaluating the academic work of colleagues and students. A clearer sense of the unavoidable ethics of turning, I suggest, pushes us to adopt a more immanent mode of critical evaluation as reviewers and evaluators of the work of our colleagues. We should enter into review processes as 'assisted reflection'. This approach to ethical reviewing is illustrated with a hypothetical example distilled from the author's actual experience.

Reflection as distancing

Pierre Bourdieu couches his approach to ethically responsible research in terms of a critique of theoretical knowledge as a way of accessing the realities of lived practical knowledge. Theory and practical knowledge, he argues, are fundamentally different:

> Because theory - the word itself says so - is a spectacle, which can only be understood from a viewpoint away from the stage on which the action is played out, the distance lies [...] not so much [...] in the gap between cultural traditions, as in the gulf between two relations to the world, one theoretical, the other practical. It is consequently associated in reality with a social distance, which has to be recognized as such and whose true principle, a difference in distance from necessity, has to be understood.
>
> (Bourdieu 1990: 14)

Researchers generate knowledge from a position characterised by 'distance from necessity' and distance from the 'urgency' of involvement in practice (Bourdieu 1990: 82). Thus they are liable to misrepresent the character and significance of practices. To avoid this, it is necessary both "to develop the theory of the logic of practice as practical participation in a game [...] and the theory of theoretical separation and the distance it presupposes and produces" (Bourdieu 1990: 104).

> Social science must not only, as objectivism would have it, break with native experience and the native representation of that experience, but also, by a second break, call into question the presuppositions inherent in the position of the 'objective' observer who, seeking to interpret practices, tends to bring into the object the principles of his relation to the object.
>
> (Bourdieu 1990: 27)

In this way "the theory of practice puts objectivist knowledge back on its feet by posing the question of the (theoretical and also social) conditions which make such knowledge possible" (Bourdieu 1977: 4).

Reflexivity and reflection have long been important themes in feminist theory. Probably the single most influential discussion of reflection appears in Donna Haraway's (1988) paper "Situated knowledges: the science question in feminism and the privilege of partial perspective". Haraway presents reflection in the context of the dilemma facing feminists and other critical scholars. On the one hand, there is clearly a need to critique traditional notions of objectivity, but on the other hand, the goal of producing "some enforceable, reliable accounts of things", or put another way, a "usable but not innocent doctrine of objectivity" remains important (Haraway 1988: 580, 582). Among the more interesting solutions, feminists have developed for this problem are 'standpoint theories' which reflectively incorporate accounts of the researcher's positionality into research projects, and 'feminist empiricism', which points out that empirical descriptions of the world are incomplete if the situations and experiences of women are either assumed identical to those of men or ignored entirely. Especially feminist standpoint theories constituted a major innovation in qualitative social science more generally.

In her 1988 paper, Haraway seeks to build upon earlier reflective approaches but to incorporate an awareness of their limitations. Her strategy is one of 'reclaiming vision', that is, seeking to recuperate a role for vision in more ethically responsible research despite the fact that visual imagery has been shown in the past to contribute to objectifying women's lives and disguising specifically masculinist ways of knowing the world as a neutral 'view from nowhere' (Rose 1993). The question is that of how to use vision but avoid a "leap out of the marked body and into a conquering gaze from nowhere", the "god trick of seeing everything from nowhere" (Haraway 1988: 581). In contrast to Bourdieu's strategy of double distancing, Haraway recommends an orientation towards the vantage points of the subjugated, who typically have less investment in remaining blind to power relations than do the privileged and powerful in any situation (Haraway 1988: 583). Vision from the bottom up, that is, can correct the distortions attendant upon the view from a privileged position. However, Haraway believes we need to perform a third reflective distancing because "[t]he standpoints of the subjugated are not 'innocent' positions" (Haraway 1988: 584). They cannot be critically effective without distancing. "One cannot 'be' either a cell or molecule – or a woman, colonized person, laborer, and so on – if one intends to see and see from these positions critically" (Haraway 1988: 585).

Accordingly Haraway seeks to fold critical distance back into the world by redefining 'rational knowledge'. In her version, rational knowledge becomes "a process of ongoing critical interpretation among 'fields' of interpreters and decoders. Rational knowledge is power-sensitive conversation" (Haraway 1988: 590). In this conversation, the object of knowledge must "be pictured as an actor and agent, not as a screen or ground or resource" (Haraway 1988: 592). In effect, by reflecting upon her own partial positionality, the researcher flattens the hierarchy implicit in the foregoing moves of distancing or abstraction, bringing herself back to a horizontal relation with research subjects. She also recognises the capacity for critical distancing in these subjects. The researcher is still separated from them, as they are from her, but not in a hierarchical, privileged way.

The geographer Gillian Rose and the philosopher and physicist Karen Barad offer interesting critiques of reflection, calling into question some of the core assumptions about distance common to Bourdieu and the Haraway of the 1988 paper. Rose's objection is to the effects of visuality. Reflection "work[s] by turning extraordinarily complex power relations into a visible and clearly ordered space that can be surveyed by the researcher: power becomes seen as a sort of landscape" (Rose 1997: 310). Such reflection does include the researcher herself in the space surveyed, but distancing remains inherent.

> Reflecting on their respective positions, a researcher situates both herself and her research subjects in the same landscape of power, which is the context of the research project in question. However, the researched must be placed in a different position from the researcher since they are separate and different from her. *Differences* between researcher and researched are imagined as *distances* in this landscape of power. [...] The researched are more central or more marginal, higher or lower, than the researcher, because they have more or less power; perhaps they are insiders while the researcher is an outsider.
>
> (Rose 1997: 312, original emphasis)

The difficulty Rose finds most telling in this practice of reflection is the fact that "(t)he researcher-self that many feminist geographers give themselves to reflect on, then, seems at some level to be a transparently knowable agent whose motivations can be fully known" (Rose 1997: 309). She finds this assumption far too optimistic.

For Karen Barad, the problem with reflection as an ethical strategy is not so much epistemological as ontological. I will dwell at more length on Barad's approach, first, because it requires somewhat more explanation, and second because it will again be taken up later in the chapter. Barad's objection to reflection is centred on the presumed ontological independence of observer and observed:

> Reflexivity takes for granted the idea that representations reflect (social or natural) reality. That is, reflexivity is based on the belief that practices of representing have no effect on the objects of investigation and that we have a kind of access to representations that we don't have to the objects themselves.
>
> (Barad 2007: 87)

Barad draws her inspiration from the pioneering quantum physicist Niels Bohr, who had disagreed with his friend and colleague Werner Heisenberg over how to interpret the fact that light could be shown to behave like a particle or like a wave depending upon experimental circumstances (Barad 2007: 106). Heisenberg had urged an epistemological explanation, according to which the physical properties of light are determinate but knowledge is inherently limited. By contrast, Bohr insisted upon a more ontological understanding of the problem:

> there aren't little things wandering aimlessly in the void that possess the complete set of properties that Newtonian physics assumes (e.g., position and momentum); rather, there is something fundamental about the

> nature of measurement interactions such that, given a particular meas-
> uring apparatus, certain properties *become determinate*, while others
> are specifically excluded.
>
> (Barad 2007: 19)

For Barad, extending Bohr's approach, this way of looking at things means that any sort of measuring or observing apparatus, whether laboratory machinery or the concepts and representations we use to grasp social and cultural reality, has to be included as part of the phenomenon we are study-ing since it is only through the 'cut' made by measuring or representing some-thing that this something becomes determinate at all (Barad 2007: 114, 146).

From such a perspective, reflection is inadequate as a way to account for the effects of research, because it assumes that the researcher and the subject or object of research, as well as the range of internal and external factors influencing the research, all existed beforehand as separate entities. "[T]he mere acknowledgment of the fact that [...] scholars are actors involved in performing their own set of practices doesn't go nearly far enough. Turning the mirror back on oneself is not the issue, and reflexivity cannot serve as a corrective here" (Barad 2007: 58). Instead of reflection, Barad recom-mends the notion of 'diffraction', or 'the production of difference patterns' first suggested in more recent writings of Donna Haraway (1997: 34, 101ff). Diffraction, too, is of course a visual metaphor. As Haraway would explain in an interview, "[v]isual metaphors are quite interesting. I am not about to give them up any more than I am about to give up democracy, sovereignty, and agency and all such polluted inheritances" (Haraway 1997: 103). What she and Barad both prefer about diffraction as against reflection is that it does not carry the implication of sameness: that what one sees in reflection is the same as the original. Diffraction, by contrast, "drops the metaphysics of identity and the metaphysics of representation and says optics is full of a whole other potent way of thinking about light, which is about [its] history" (Haraway 1997: 103).

In physical terms, diffraction "has to do with the way waves combine when they overlap and the apparent bending and spreading of waves that occurs when waves encounter an obstruction" (Barad 2007: 74). Unlike par-ticles, Barad goes on to explain, "waves can overlap at the same point in space. When this happens, their amplitudes combine to form a composite waveform. For example, when two water waves overlap, the resultant wave can be larger or smaller than either of the component waves. [...] Hence the resultant wave is a sum of the effects of each individual component wave" (Barad 2007: 76). If the medium of reflection is the mirror, that of diffraction is the 'diffraction grating', by which Barad means "an apparatus or mate-rial configuration that gives rise to a superposition of waves" (Barad 2007: 81). She interprets diffraction gratings very broadly to include, for example, texts in which different lines of thinking are brought together and 'interfere' with each other in ways productive of new differences.

To translate this perspective into the terms of social or cultural research practices, the perspective of the researcher should not be seen to hover above or at a distance from that which she interprets (including her own position). Instead, the effects of the researcher are 'interferences' operating at the same level as what is being studied. Further, what is being studied itself comes to appear as such only through the specific kind of 'cut' the researcher makes between 'observer' and 'observed' with the help of specific categories, distinctions and naming practices. Thus, diffraction is not about distance or separation but entanglement, another term from quantum theory that describes the co-variation of physical states – for example of particles at a great distance from each other – which appear to have no way of influencing each other. In Barad's view, we are always already in the midst of processes of becoming. "We don't have the distances of space, time and matter required to replicate 'what is' [as in a mirror]; in an important sense, we are already materially entangled across space and time with the diffractive apparatuses that iteratively rework the 'objects' that 'we' 'study'" (Barad 2007: 384). Barad proposes that social scientists, science studies scholars and others should practice diffractive reading and thinking, that is, "thinking insights from different disciplines (and interdisciplinary approaches) through one another" in a way that "does not take the boundaries of any of the objects or subjects of these studies for granted but rather investigates the material-discursive boundary-making practices that produce 'objects' and 'subjects' and other differences" (Barad 2007: 93). In sum,

> a diffractive methodology is respectful of the entanglement of ideas and other materials in ways that reflexive methodologies are not. In particular, what is needed is a method attuned to the entanglement of the apparatuses of production, one that enables genealogical analyses of how boundaries are produced rather than presuming sets of well-worn binaries in advance.
> (Barad 2007: 29–30)

Barad closes her argument by drawing upon Emmanuel Levinas, who places a primordial ethical responsibility at the core of what it is to be a subject. "We (but not only 'we humans') are always already responsible to the others with whom or which we are entangled, not through conscious intent but through the various ontological entanglements that materiality entails" (Barad 2007: 393). Entanglement and the new forms of ethical responsibility associated with it, are thus fundamentally about the lack of distance separating researcher from researched. Diffraction is, in a sense, about how to conceive of ethics without distance.

Turning and attention

The larger intent here is to critique both the understandings of reflection in terms of distance in Bourdieu and the earlier Haraway and also the critiques of reflection offered by Rose and Barad. But before offering a critique of these positions, it is worth preparing the ground by turning briefly to a discourse that has emerged in the last two decades around the economics and politics of attention. This discourse is helpful because it re-frames political and ethical questions not in terms of distance at all but rather in terms of direction. The starting point for much of this discourse is what might be called our 'directional finitude', our limited capacity of being turned towards things, the fact that there is always far more 'behind our backs' than 'in front of' us (Hannah 2015; Hannah 2019). It is not an accident that an ethical and political concern with our limited directional capacities has emerged in recent discussions of 'attention economies' (Ash 2012; Beller 2006; Crary 1999; Goldhaber 1997; Franck 1998; Stiegler 2010; Wilson 2015; Wu 2016). Attention, as treated in the psychological and phenomenological literature, designates a highly complex and multi-faceted set of phenomena (Styles 2006; Waldenfels 2004; Wehrle 2013). It involves not just what we normally think of as attention, namely focused, cognitive concentration on something, but also a range of other phenomena involving bodily and affective capacities and processes that are often neither focused nor deliberate in character. Nevertheless, humans obviously relate to the world in important ways through attention in the classic sense of relatively focused engagement with some matter of concern. To a significant extent, our daily activities, whether at work, at home or elsewhere, can be characterised as a sequence of directed engagements. As psychological research continues to show, focused engagement of our attention is inherently selective, all popular talk of 'multi-tasking' aside (Gazzaley and Rosen 2016). It is this feature of attention that plays a central role in recent writings.

As the 'Information Age' has deepened its hold on forms of communication and calculation, it has become increasingly evident that the most valuable commodity in capitalist economies is not information per se but attention, or more broadly, our ability to be engaged with any particular piece of information. Herbert Simon already pointed out in 1971 that a valuable commodity is something scarce, and it is not information that is now scarce but rather our ability to be focused upon it, that is, human attention (Simon 1971). Especially relatively privileged populations of the Global North face an increasing disproportion between the exploding range of possible focused engagements of our attention and our extremely limited and only modestly expandable ability to engage with any particular matter of concern (Gazzaley and Rosen 2016). It is not accidental that some of the most influential accounts of the attention economy have been modelled upon academic life (Goldhaber 1997; Franck 1998). It is arguably in academic life that the commodification of attention has attained some of its most refined expressions, in the form of citation indices and impact factors. Attention can be measured in these indices in a way that has nothing do with qualitative aspects of the work: the mere fact of having been cited already indicates that some colleagues engaged in a particular debate feel it necessary to attend to a particular study (Franck 1998). In sum, the attention of colleagues to one's own work is a kind of currency we can accumulate.

Academic life is suffused with an attention economy in more collective senses as well. The sequence of 'turns' that have become ever more frequent as part of the larger 'cultural turn' of the 1990s carries this implication (Bachmann-Medick 2006). 'Turns' are often presented as collective decisions to drop a paradigm or general line of research in favour of alternative directions, although many such turns, such as the 'spatial turn', the 'practice turn' (Knorr Cetina et al. 2001), or any of a number of other such turns, are better described as branchings that open up new possible arenas of research alongside the ongoing pursuit of pre-existing lines of work (Bachmann-Medick 2006). Nevertheless, for individual scholars, the announcement of a new 'turn' raises the inherently ethical issue of how much time and effort to devote to the new direction in view of whatever one was already doing. Though not the main topic of the present chapter, this is itself an under-appreciated ethical quandary: how much does one need to know about a new research direction in order to decide in a responsible and defensible way whether one needs to know more about it? At a smaller and more temporary scale than that of 'turns', the seemingly endless string of 'calls' for papers at myriad conferences invokes an ethics of directed attention. How many scholars will respond to any given call? Will it be possible to stage multiple sessions? How many will turn up to listen during the conference? The presupposition behind this economy of CfPs is precisely the scarcity of attention characterising the larger academic attention economy.

What is crucially important to take from this discourse on attention is that the political and economic (and likewise the ethical) dimensions of turning towards one matter of concern are *different* from the kind of substantive ethics with which we often concern ourselves. Beneath or prior to the issue of exactly what we think, feel, or do in relation to something is *the question of whether we are turned toward it or away from it*. If we are turned towards something, whether an activity, a thought, another person, an object, or a relation, it is possible for us to engage with it actively, creatively, critically, etc. If we are turned away from it, not engaged with it, it is inert for us, not currently animated. Critiques of the cultural productions of 20th- and 21st-century capitalism have long pointed to the distracting and de-politicising effects of mass distraction (Benjamin 1989) and spectacle (Debord 2010). Recent critical work on attention can be seen as a continuation and refinement of these earlier insights. Equipped with a stronger sense of the ethics and politics of directed engagement highlighted by work on the attention economy, we can now return to the ethics of reflection.

Reflection as distancing and turning

The notion of reflection as treated by Bourdieu and Haraway, but also the critique of reflection by Rose and Barad, are based upon an incomplete account of reflection. Reflection, as the visual and optical metaphor should already strongly suggest, is not merely a distancing but at the same time a *turning back* upon one's own perspective. Turning back to reflect presupposes distancing, but is not reducible to it. The English 'reflect' derives from the Latin *reflectere*, 'to bend back' (Merriam-Webster 1981: 963). This bending or turning back itself has important effects of separation *different from, and independent of,* the separation inherent in distancing. This can be seen if we look more closely at the arguments about reflection from Bourdieu, Haraway and Barad. If we start with Bourdieu, we can begin to get a sense for how reflection as turning can be understood. Bourdieu's discussion of reflecting upon theoretical distance suggests a linear geometry: we compound the first distancing inherent in theory itself with a second distancing in the form of a 'theory of theoretical practice'. But the second distancing is always also a turning-back, and thus automatically comes with opportunity costs, matters dropped or left aside in order to reflect. We can see this if we follow what Bourdieu actually does in the two major books laying out his theory of practice. In both cases, he first turns to theoretical practice as his object, and then, equipped with a better sense of how theory performs distancing, he turns to the logic of practice (Bourdieu 1977, 1990). In an extremely banal sense, the term 'turning' is appropriate here because it carries with it the connotation of exclusivity. Bourdieu cannot both theorise about theoretical practice and simultaneously theorise about non-theoretical practice. To the extent that he does the latter, he postpones doing the former, and vice versa. The relation between primary research and reflection involves an OR; it is not simply an AND-relation. The same is true of the feminist ethics of reflexivity. Reflective self-positioning is an activity of thought or writing that precludes other activities. When we turn or 'bend back' to reflect upon our positions, we are not turned towards other things. Here, too, we are encouraged to assume that the results of reflection in these two cases are somehow *added to* the first-order insights of our research without exacting a price, that reflection upon research is a simple enhancement of its validity. But there is always an opportunity cost to reflection. This applies at the collective level of academic work as well as at the individual level.

If the relation of reflection to the research reflected upon is not just an additive AND relation but in part also an OR relation, this suggests a complication of the traditional view of the ethical function of reflection. This traditional, additive view is often expressed in terms of a 'failure to include', as when Bourdieu writes that "any analysis of ideologies [...] which fails to include an analysis of corresponding institutional mechanisms is liable to be no more than a contribution to the efficacy of these ideologies" (Bourdieu 1977: 188). This kind of admonishment is ubiquitous in critical social research: 'failure to include' an institutional analysis, 'forgetting' to discuss the political function of ideologies, 'reproducing' 'oversights and omissions' through 'complicitous silences': all of these formulations presuppose that it is practically possible to avoid all such silences, omissions and oversights in carrying out and publishing research. Bourdieu in effect demands comprehensive reflection in addition to first-order research. This ignores the

OR-character of the relation between research and reflection upon it. At the very least, the standard for ethical research cannot be one of merely requiring the maximum possible amount of reflection. At least in part, it must involve weighing priorities between and among different aspects of the research itself and reflections upon it.

Does Karen Barad's replacement of reflection with 'diffraction' help overcome this problem? Barad claims that what we think of as reflection is actually not available to us: '[w]e don't have the distances of space, time and matter required to replicate 'what is' [as in a mirror]; in an important sense, we are already materially entangled across space and time with the diffractive apparatuses that iteratively rework the 'objects' that 'we' 'study' (Barad 2007: 384). In other words, distancing is not actually possible. And yet Haraway's point that "(o)ne cannot 'be' either a cell or molecule – or a woman, colonized person, laborer, and so on – if one intends to see and see from these positions critically" (Haraway 1988: 585) remains valid as well for Barad. A certain reflective turning is hard to eliminate. This is most easily visible in Barad's reliance upon diagrams to explain superposition and interference, entanglement, complementarity and other concepts central to her argument. The images, which are supposed to help demonstrate that the distance necessary for reflection cannot be attained, *can themselves only be comprehended through a 'view from above'*, that is, from a distance. And this distancing, because it is also a turning, temporarily rules out practical involvement in what is being pictured. This way of approaching the problem of reflection should not be foreign to Barad's argument: after all, it could be seen as an analogue of the very problem that kicked off quantum theory: the impossibility of measuring both position and momentum of a particle simultaneously. The contention here is that we should see primarily academic analysis and reflection upon it as likewise mutually exclusive. Barad seems in places to acknowledge the unavoidable turning that haunts her account of diffraction: "at times diffraction phenomena will be an object of investigation and at other times it [*sic*] will serve as an apparatus of investigation; it cannot serve both purposes simultaneously since they are mutually exclusive" (Barad 2007: 73). In other words, turning back to consider diffraction is not the same as 'simply' diffracting. Put differently, insofar as diffraction is thematised *at all*, it becomes 'polluted' with reflection, and thus with the exclusive economy of turning. In short, Barad, too, must engage in reflection, and thus must deal with the OR-relation inherent in turning away from one thing to address another.

Despite this, Barad exhorts her readers to comprehensive assumption of ethical responsibility in a way very similar to Bourdieu. Speaking of reflection, she argues that 'one can't simply bracket (or ignore) certain issues without taking responsibility and being accountable for the constitutive effects of these exclusions' (Barad 2007: 58). Later in her argument, she approvingly cites the "political potential of deconstructive analysis", which "lies not in simply recognizing the inevitability of exclusions but in insisting on accountability for the particular exclusions that are enacted and in taking up the responsibility to perpetually contest and rework the boundaries" (Barad 2007: 205). Haraway, too, in her later discussions of diffraction, obscures the inherent limitations in our ability to turn to different things simultaneously. She recounts to Thyrza Nichols Goodeve the way she used a diaper pin worn by a student on her hat as an occasion for diffractive think-ing. The student was a politically engaged midwife for whom the diaper pin symbolised her opposition to medically mediated childbirth. Haraway 'diffracted' the diaper pin by leading the students through a discussion of its place in the history of the plastics and steel industries, the history of safety regulation and of capitalism more broadly. This diffractive exercise, she says, was meant "to show that [the diaper pin] has many more meanings and contexts to it and that once you've noted them you can't just drop them. You have to register the 'interference'" (Haraway 1997: 105).

Barad's admonishment to "perpetually contest and rework the bounda-ries" amounts to a call 'perpetually' not to attend to anything else. Likewise, Haraway's ethical claim that once additional contexts for something have been recognised, 'you can't just drop them' presupposes an unrealistic, arbi-trarily expandable ability to be turned towards ever more matters of con-cern. Reflection upon something always means being turned away from other matters, and this, in turn, means *stabilising these other matters by not relating to them.* In not relating to something, we are not actively stabilis-ing it, for example, by naming it. On the other hand, we are also not de-stabilising it or treating it as unstable, contingent, or contestable. We are not directing our critical attention to it, even if, at other times, we may do so, and in doing so, may construct or perceive it as unstable and contingent. Put differently, *instability, not just stability, is performative.* Our inability to avoid reifying, essentialising, objectifying or otherwise stabilising at least some matters at all times is not a question of insufficient epistemological 'radical-ity' but rather a result of the fact that all possible epistemological positions are circumscribed in a practical sense by our inability to turn our attention and practice critically towards all things at once. Thus the ethics in research cannot consist of the elimination of all distance. Nor can it be to attain a state in which one is 'turned critically toward' every aspect of research simul-taneously. The ineliminable effect of our directional finitude is that there is *always* some aspect of our research that we are not actively de-stabilising.

Turning and the ethics of academic reviewing

The foregoing is intended at least to suggest how our perspective on aca-demic life can be altered by sensitivity to dynamics of turning towards and away from different matters of professional concern. Here I would like to focus, albeit also in a fairly brief way, on the practice of reviewing or evalu-ating the work of other academics, whether in the form of peer reviews, book reviews, grant evaluations or related exercises. Reviewing is an appropriate focus here because the way we review academic work by others says a lot about how we think it should be carried out in general. Reviewing is an expression of our own otherwise implicit understandings of our profession. It is useful to focus upon reviewing also because of the disproportionate relationship between its crucial importance to the production of knowledge and its low status as an institutionally recognised activity. The exploitative character of reviewing as it is currently organised has been noted countless times, but I think we can understand this situation also as an expression of a generalised lack of recognition of the finite selectivity of our directed engagements. Directional ethics of academic practice would accord reviewing activities the same status as all other directed activities.

The second basic point to be made here builds upon the considerations discussed above. I would like to suggest that the review process is best thought of as what might be called 'assisted reflection'. As noted at the beginning of this chapter, one of the two main functions of systematic reflection upon research is to improve it through rendering it as transparent as possible. Peer review has the closely related function of ensuring the quality of published research through procedures that render it transparent, with the difference, of course, that in the latter case, the research is not made transparent through reflection by the researchers themselves but rather through exposing it to external scrutiny. Sound research as submitted to journals or funding agencies in the social sciences and cultural disciplines is expected already to include evidence of systematic reflection by the author(s). External reviewers can then be understood to provide additional perspectives. My contention is that ethically sound reviewing should be understood by reviewers themselves and by editors as second-order reflection, as a continuation of reflection, or, as suggested, as 'assisted reflection' rather than as a fundamentally different form of transparency. I mean two things by this. First, reviewers should be expected to place themselves as far as possible within the perspective of the original authors, and to approach the strengths and weaknesses of a particular submitted piece of work from within the priorities set by the authors. This is what has long been known as 'immanent critique'. Immanent critique seeks to evaluate an argument as far as possible according to *its own standards*, not against standards or criteria imported from other perspectives. This is an ethically sound approach in the sense that it acknowledges that priorities must be set, and refrains as far as possible from doing violence to the specific priorities set by authors.

Immanent critique is not itself a new approach. However, I want to argue that an understanding of the finitude of attention, and of the turning involved in reflection, provide a new, additional reason to pursue immanent critique. Being turned towards research in reflection excludes being turned towards the objects of research within the research itself, and vice versa. More generally, being turned towards any particular facet of research precludes being turned towards other facets. Authors, in the 'urgency' of academic practice, trying to publish the results of research, must treat reflection upon and reporting of research, as well as all different possible emphases within the reporting of research, in terms of an OR, that is, as mutually exclusive things to pursue in the context of writing a paper or a proposal, as a zero-sum game. The imperative to weigh the relative importance of reflection and research, or to prioritise specific elements at the expense of others within either of these categories, is constant. This imperative faced by authors should be fully adopted by reviewers. Not only should reviewers seek to judge and criticise as far as possible on the basis of the criteria set out by the authors, but they should also constantly hold in view the fact that any suggested additions (of topics, discussions of other sources, etc.) come at the expense of some other aspect of the work. This is the second part of what I mean by reviewing as 'assisted reflection'. A number of other ethical problems encountered in reviewing, for example, harsh rhetoric, attempts to guess the real author, or the privilege that comes from having read ground-breaking work before others have had a chance to read it, might all be alleviated by a stronger focus upon immanent critique. This is because immanent critique, in urging reviewers to adopt the perspective of authors to the maximum extent possible, tends to encourage identification with the authors' project rather than the assumption of a hostile or competitive attitude towards it.

The impacts a more directional perspective could have can be illustrated with a fairly simple, hypothetical example. Reviewers come from outside a piece of research and are easily tempted to suggest that authors consult additional literature. This is, to be fair, one very common way in which reviewers are able to feel useful to authors and editors. Reviews that do not suggest any substantial additions to the literature consulted are quite rare, in my experience. A paper on refugee camps, for example, that describes them as 'spaces of exception' will be expected by most reviewers to include some discussion of the work of Giorgio Agamben. The ethical question is, to what extent should the author(s) be held responsible for an account – and for what kind of account – of Agamben's writings on states and spaces of exception? In my experience, it is all too common for reviewers who know the relevant theoretical literature well to urge the inclusion of additional writings from this literature. The better they know this literature, the more likely it is that they will recommend substantial additions from it. A stronger awareness of the ethics imposed by the economy of attention should give pause here. It should lead reviewers to acknowledge the inescapability of opportunity costs to whatever they might suggest. Their first question should be about the purpose of the paper as the author(s) conceive it. If the main point of the paper is empirical, the default assumption should be that no additional material on Agamben is necessary. Of course, it could be that the specific way Agamben is summarised, or the specific works serving as the basis for the summary, need improvement. But again, a simple expansion of the treatment of Agamben should not be urged. If, by contrast, the main point of the paper is to use the empirical example to suggest changes in how we understand the concept of spaces of exception, a detailed and differentiated theoretical discussion is to be expected. Reviewers should in this case refrain from urging the inclusion of more empirical material except insofar as they can make a convincing argument that it would substantially assist the author(s)' primarily theoretical project. In both cases, any suggested addition to an argument should be accompanied by suggestions about what material can be cut out or handled more briefly. The reviewer, in other words, should fully inhabit the economy of finite attention which the author has no choice but to inhabit.

Documents outlining reviewing procedures are provided by all reputable academic journals and funding agencies, and these documents usually explicitly address ethical issues. The Committee on Publication Ethics (COPE) provides online resources covering, among other areas, the ethics of reviewing (Hames 2013). COPE's guidelines for ethical reviewing are more comprehensive and detailed than those of most journals, so they can serve as an illustrative case here. Some points appearing in these guidelines at least implicitly gesture towards the issues raised here. Under the heading "When preparing the report", COPE's guidelines urge reviewers to "remember it is the authors' paper and not attempt to rewrite it to their own preferred style if it is basically sound and clear" (Hames 2013). From the perspective put forward in the present chapter, this principle should be formulated to address not just style but the prioritisation of substance: the fact that "it is the authors' paper" should be respected not just in how it is written but in the chosen topic and the general shape of the argument. COPE's guidelines also at least faintly acknowledge the difficulties imposed upon authors by recommendations that they include more material in their written work. Under the same heading as the previous point, COPE's guidelines urge reviewers to "make clear which suggested additional investigations are essential to support claims made in the manuscript under consideration and which will just strengthen or extend the work" (Hames 2013). The implication seems to be that in the latter case, where suggested additions would "just strengthen or extend the work", there is some degree of freedom as to whether the additions must actually be taken on. In the perspective sketched here, this faint signal must be strengthened: especially suggestions that urge authors to 'extend' their published arguments should carry with them the burden of proof that the suggested extensions are more important, *by the authors' own criteria*, than some other segment of the work that would have to be removed to make room for the suggested extension. Editors should be vigilant on this point and seek to hold reviewers to the relatively strict substitutional ethics required by the finitude of attention.

Conclusions

This chapter has sought to argue for a different way of understanding the ethics of reflection in academic work and to illustrate how this different perspective could shift typical practices of academic reviewing. The ethical purpose of reflection was elaborated through summaries of the work of Bourdieu and Haraway on reflection and of critiques of reflection by Rose and Barad. It was argued that all of these approaches to reflection fail adequately to acknowledge the inherently exclusive moment of turning at work in all reflection. A brief summary of recent writings on the politics and economics of attention was provided in order to sharpen the sense of the exclusiveness of our engagements based on our inherent directional finitude. Equipped with this sensitivity, it was shown that if we attend to what Bourdieu and Barad actually do in presenting their arguments about reflection, it is clear that a moment of exclusive turning is unavoidable, and that this implies a different kind of ethics than the usual substantive ethics. This different ethics operates at the more basic level of what we are turned towards and away from, regardless of what we specifically think, say, write or do while being turned towards something. In the final section, a more directionally sensitised perspective on academic work was mobilised to suggest a different approach to the ethics of academic reviewing. Academic reviewing is an insufficiently acknowledged microcosm and an expression of how academics understand the way academic work should be done. Reviewers, it was argued, should be obligated to inhabit the directional finitude of authors as far as possible, and to approach reviewing as 'assisted reflection' that takes into account directional finitude. Doing so is more respectful of the (inherently exclusive) priorities set by authors themselves, and tends as well to address many of the other ethical problems associated with reviewing.

References

Ash, J. (2012) 'Attention, Videogames and the Retentional Economies of Affective Amplification', *Theory, Culture & Society*, 29 (6): 3–26.

Bachmann-Medick, D. (2006) *Cultural Turns: Neuorientierungen in den Kulturwissenschaften*, Reinbek bei Hamburg: Rowohlt.

Barad, K. (2007) *Meeting the Universe Halfway: Quantum Physics and the Entanglement of Matter and Meaning*, Durham, NC: Duke University Press.

Beller, J. (2006) 'The Cinematic Mode of Production: Attention Economy and the Society of the Spectacle', *Journal of Communication Inquiry*, 32 (2): 204–209.

Benjamin, W. (1989) *Das Kunstwerk im Zeitalter seiner technischen Reproduzierbarkeit*, Stuttgart: Reclam.

Bourdieu, P. (1977) *Outline of a Theory of Practice*, Cambridge: University Press.

Bourdieu, P. (1990) *The Logic of Practice*, Cambridge: Polity Press.

Crary, J. (1999) *Suspensions of Perception: Attention, Spectacle, and Modern Culture*, Cambridge, MA: The MIT Press.

Debord, G. (2010) *Society of the Spectacle*, Eastbourne: Soul Bay Press.

Franck, G. (1998) *Ökonomie der Aufmerksamkeit: Ein Entwurf*, 11th edition, Munich: Carl Hanser Verlag.

Gazzaley, A. and Rosen, L. (2016) *Distracted Mind: Ancient Brains in a High-Tech World*, Cambridge, MA: The MIT Press.

Goldhaber, M. (1997) 'The Attention Economy and the Net', *First Monday*, Volume 2 (4-7), https://firstmonday.org/article/view/519/440#dep6 (accessed 13 January 2019).

Hames, I. (2013) COPE Ethical Guidelines for Peer Reviewers, http://publicationethics.org/files/u7140/Peer%20review%20guidelines.pdf (accessed 13 January 2019).

Hannah, M. (2015) 'Aufmerksamkeit und geographische Praxis', *Geographische Zeitschrift*, 103 (3): 131–150.

Hannah, M. (2019) Direction and Sociospatial Theory: A Political Economy of Oriented Practice, London: Routledge.

Haraway, D. (1988) 'Situated Knowledges: The Science Question in Feminism and the Privilege of Partial Perspective', *Feminist Studies*, 14 (3): 575–599.

Haraway, D. (1997) *Modest_Witness@Second_Millennium. FemaleMan_Meets_OncoMouse™*, London: Routledge.

Knorr Cetina, K., Schatzki, T. and von Savigny, E. (eds) (2000) *The Practice Turn in Contemporary Theory*. London: Routledge.

Merriam-Webster (1981) *Webster's New Collegiate Dictionary*, Springfield, MA: Merriam & Co.

Rose, G. (1993) *Feminism and Geography: The Limits of Geographical Knowledge*, Cambridge: Polity Press.

Rose, G. (1997) 'Situating Knowledges: Positionality, Reflexivities and Other Tactics', *Progress in Human Geography*, 21 (3): 305–320.

Simon, H. (1971) 'Designing Organizations for an Information-Rich World' in M. Greenberger, (ed) *Computers, Communication and the Public Interest*, Baltimore, MD: The Johns Hopkins Press, pp. 38-72.

Stiegler, B. (2010) *Taking Care of Youth and the Generations*, Stanford, CA: Stanford University Press.

Styles, E. (2006) *The Psychology of Attention*, 2nd edition, New York: Psychology Press.

Waldenfels, B. (2004) *Phänomenologie der Aufmerksamkeit*, Frankfurt am Main: Suhrkamp.

Wehrle, M. (2013) *Entwurf einer dynamischen Konzeption der Aufmerksamkeit aus phänomenologischer und kognitionspsychologischer Sicht*, Paderborn: Wilhelm Fink.

Wilson, M. (2015) 'Paying Attention, Digital Media, and Community-Based Critical GIS', *Cultural Geographies*, 22 (1): 177–191.

Wu, T. (2016) *The Attention Merchants: The Epic Scramble to Get Inside Our Heads*, New York: Alfred A. Knopf.

Index

Aboriginal communities 8–9
Aboriginal research ethics 9
abuse 3; child 71–74, 116; contact sexual 71; domestic child 61, 71–73; non-contact sexual 71
academic communities 55
academic productivity 175
academic reviewing 234–237, 238
activism research 136
affect 146
affective 157–158; capacities and processes 230; cultural practices 172; phenomena 214
American Association of Geographers (AAG) 30, 53
American Psychological Association 24
anonymity 28, 82, 84, 114, 122–124; declining 161; and 'ethics of recognition' 137; physical 85; of research participants 118
Antwerp 124, 126
approaches to ethics didactics 204–207
Askins, K. 10, 137
assisted reflection 224, 235–236, 238
attention and turning 229–231
attention economies 5, 230–231, 236–238
Australia 69, 168, 173
Australian National Statement on Research Ethics 4
autoethnography 29
autonomy: children's 65; highly valued research 31; of Indigenous peoples 25; individual 29; researcher 137

Barad, K. 224, 226–229, 231–234, 238
Barnett, C. 6, 47
Baumann, C. 6, 14, 214

Belarus 82–83; Afghan refugee and asylum-seeking women in 85; practising research with refugees and asylum seekers in 82–83; qualitative fieldwork in 79
Belmont Report 29
beneficence 29, 31, 140–142, 168
Beurskens, K. 8, 100–104
Blazek, M. 10, 92, 101, 134–135, 139, 146, 177
Bourdieu, P. 224–226, 229, 231–233, 238
Bowman expedition 25
Bruns, B. 9, 79, 84, 102, 118, 121
burnout 115, 175; emotional 170

care 143–144; about research ethics 33; care at-distance 146; duties of 11, 65, 69, 73, 83, 174; ethics of 11, 144, 175, 176–177; ethos of 170; feminist ethics of 11, 175; politics of 139–140, 145–147, 177
caring: about research ethics and integrity 23–36; professions 135–136, 138–139
caring professions: ethics in the helping and 138–139; geographical research and helping and 135–136
case studies: conducting research on diamond traders 124–128; mapping Muslim neighbourhoods 186–190; release of data 190–196; researching cross-border cigarette smugglers 121–124
child abuse 71–74, 116; domestic 71–73
childhood: conceptualisations of 73; contact sexual abuse during 71; diversity 71; as a foreign country 59–75; memories of 74; non-contact sexual abuse during 71

children: competencies 65; consent 66–70; decision-making rights 61; participation rights of 61; rights to protection from harm 59; wellbeing 74

children's geographies 66, 71, 74, 116

choices: ethical 105, 129; methodological and epistemological 5

cigarette smugglers 115, 120, 121–124

classrooms, doing geography in 213–221

closed contexts 115, 117–118, 120

codes of conduct 23, 138, 142

codes of ethics: formal 53; general 73; institutional 28

collaborative knowledge production 98

collaborative research 8

coloniality of knowledge production 93, 98, 106, 108

Committee on Publication Ethics (COPE) 33, 237

communal relationships 29

communities: aboriginal 8–9; academic 55; caring research 177; economically and socially marginalised 59; Inuit 49; local 34, 49–50, 121–122; marginalised 107; Muslim 186–189; Pacific Island 31; public 183; refugee 83–84; religious 125; scientific 2, 42, 183; socially underprivileged 169; trader 126

community-based adaptation 49–51; and ecosystem management 49–51

community-based charity organisation 134

community-based organisations 146

community-based participatory research 8

community-based research 8–9, 49–51

community-minded organisations 32

community-related knowledge production 9

conditions of knowledge production 9–10

confidentiality 12, 28, 68, 129, 137–139; breaching 118; of data 118; principles of 74; research 84; research participant 161

conflict: between diverse ethical pressures 108; ethical 52, 54; of interest and roles 25, 173; military 203; political 203

consent 66–70

consequentialism 29

contact sexual abuse 71

contexts of knowledge production and application 104–108, 202–203

contextual sensitivity 13

covert research 29

criminal law 120

cross-border cigarette smugglers 121–124; appropriate methodological framework 122–123; defining sensitive character of the study 121–122; protecting research subjects' and researchers' rights 123–124

cultural insider 83–84, 87

data handling and research ethics 12–13

deception 85, 197

decolonial thinking 98–99; post-Soviet 93

Dekeyser, T. 5, 155

deontological ethics 29, 64

descriptive ethics 7, 45

de-Westernisation 93–94, 108

diamond trade 115, 120, 124–128

diamond traders 124–128; appropriate methodological framework 125–127; defining sensitive character of study 124–125; protecting research subjects' and researchers' rights 127–128

Dickel, M. 14

diffraction 228–229, 233–234

direction 188, 230–231; of the event 158; of future research 43; research 231; of work 51

directional ethics 223–238

disasters 169, 172

disciplinary debates on research ethics in human geography 6–7

distance/distancing 225–229, 231–234

distress 25, 69, 170; emotional 115; mental 25; physical 115; psychological 115

distributive justice 29

diversified methodologies 70–71

doing geography in classrooms 213–221

domestic child abuse 61, 71–73

Dominey-Howes, D. 11, 169–172, 175, 177

'Do No Harm' 81, 85, 168

Drozdzewski, D. 11, 169–172, 174–175, 177

duties of care 11, 65, 69, 73, 83, 174

Eastern Europe: inequalities of knowledge production in 92–109; and research ethics 92–109 economically marginalised communities 59

ecosystem management: community-based adaptation and 49–51; and policy recommendations 51–52

'edge ethnography' 155

education: research ethics debates and 29–30; scientific 14

educational philosophy 213–214

emotional aspects 101

emotional burnout 170

emotional distress 115

emotional labour 145, 175

emotional wellbeing 169

emotions 66, 101, 115, 211

empathic engagement 171

entanglement 233; and diffraction 229; ethnographic 155; ontological 229

epistemological: bias 99–100, 173, 176; choices 5; deadlock 153; explanation 227; foundations 108; inconsistency 99; interests 8; limitations 223; positions 223, 234; 'radicality' 234; shift 203

epistemology biases 99–100, 173, 176

ethical: beliefs and assumptions 51; calamities 23; coding 156; decision-making 6; dilemmas 30, 42, 47, 50, 66, 74, 79, 86, 105, 124, 127, 160, 163; dimension of teaching and learning 213–221; philosophies 134; principles 3, 29, 42, 73, 138–139, 201–202, 204, 206; protocols 63, 137; reasoning 1, 6–7, 9, 14, 52; reflexivity 65; scientific standards 42; sensitivity 10; standards in research 3, 5, 8, 53–55, 106, 203; symmetry 63

ethical challenges: arising from vulnerability of asylum seekers 78–87; arising from vulnerability of refugees 78–87

ethical dilemma 30, 42, 47, 50, 66, 74, 79, 86, 105, 124, 127, 160, 163

ethical dimension of teaching and learning 213–221

ethical judgement 163

ethical literacy 44, 54–55

ethical norms 14

ethical protocols 63, 137

ethical research principles: teaching to geography students in higher education 200–211

ethical responsibility 138, 162, 203, 229, 233

ethical sensitivity 10

ethical values 139

ethics: deontological 29, 64; descriptive 7, 45; directional 223–238; helping and caring professions 138–139; institutional 5, 7, 30, 35, 106; institutionalised 2–5, 47; and institutional requirements 13–14; normative 7, 11, 207; practical 2, 182–197; procedural 2, 12–13, 23, 137; relational 137–138; in socio-spatial childhood research 59–75; and speed of research practice in neoliberal university 173–176; teachable 204–207

Ethics, Place and Environment 6

ethics committee 4, 13, 34, 53, 156–157, 160, 162, 171, 177, 185

ethics creep 174–175, 177

ethics didactics, approaches to 204–207

ethics governance 31

ethics of academic reviewing and turning 234–237

ethics of care 11, 144, 175, 176–177; writing researcher trauma into 176–177

ethics of intervention 51–52

ethics of reflection 223–238; overview 223–224; reflection as distancing 225–229; reflection as distancing and turning 231–234; turning and attention 229–231; turning and ethics of academic reviewing 234–237

ethics panel 173

ethics review: institutional 35; and researcher trauma 177

ethics review processes 55, 169, 173–174, 176

ethnographic entanglements 155

ethnographic fieldwork 104, 122, 163, 199

ethnographic qualitative methods 78

ethnography 136, 163

ethos of care 170

Eurasia: inequalities of knowledge production in 92–109; and research ethics 92–109

evaluation 45, 116; critical 197; of ecosystem management 51; ethical 155; of human action 204; immanent 160; moral 205; of research ethics 45; of soil erosion and land use maps 208

experience: Eastern Bloc 93; fieldwork 93, 108, 135; process 220; in the research field and in academia 96–100; structure of 218; traumatic

170; as volunteer-practitioners 147;
Western Bloc 93
experience-oriented geography
lessons: experience process 220;
facets of 218–220; giving topic a
subject-specific perspective 219;
selecting viable example 218–219;
tactful interaction between pupils
and subject matter 219–220
explorative learning 213–214
exposure 9, 84, 125, 170, 174
external regulation 52–53;
self-regulation and 53, **53**
Eyles, J. 6

fabrication 31
facets of experience-oriented
geography lessons 218–220
falsification 31, 47
feminist 8; empiricism 226; ethics of
reflexivity 232; geographers 78, 81,
227; geography 172; philosophy 137,
140; scholarship 106
feminist ethics of care 11, 175
feminist praxis 169, 171–173
fieldwork relations 10–11, 93–94
First World War 203
formal codes of ethics 53
friendliness 144
friendship 97, 126, 128, 143–145
friendship-liness 133–147, 142–145

Garrett, B. 5, 27, 153n2, 155
gatekeepers and vulnerability 83–85
GDR 93, 100
geographical education 219–221
geographical fieldwork 60
geographical practices 200–201
geographical research and helping and
caring professions 135–136
geographical science and
representation 202–204
geographies: moral 6; and relational
ethics 137–138
geography lessons: activity field
of multiple decisions 215–216;
ethical dimensions 216–217;
experience-oriented 218–220; facets
of experience-oriented geography
lessons 218–220; subject-specific
understanding of topic 216–217;
transferring subject-specific
understanding 217
Georgia 93, 96–99, 104–105, 107–108

geospatial technologies 28
GIS 14, 47, 182, 186
GIS Professional Ethics project 197
globalisation 42, 215
Gregory, Derek 203
group discussions 82–84, 100–104, 197

Hannah, M. 11, 230
Happ, D. 9, 13, 85
Haraway, D. 224–226, 228–229,
231–234, 238
harm 1–2, 11, 23; avoidance of 139;
minimise 24–27, 71–72; physical 85;
psychological 85; social 85
Harvey, F. 14, 183
Hay, I. 5–7, 23–25, 30, 32, 47, 54
helping and caring professions: ethics
in 138–139; geographical research
and 135–136
Henn, S. 9, 44n1, 118, 124
hermeneutics 216–217;
phenomenological 215
hidden populations 117
higher education: institutions 176, 211;
teaching ethical research principles
to geography students in 200–211
historical 3, 9, 30; analyses 47; contexts
of knowledge production 202–203;
geographies of motherhood 116;
phases and societies 206
Hörschelmann, K. 11, 13, 68, 71, 92–94
human geographers and research ethics
44–46
human geographical research ethics 14–15
human geography: caring about
research ethics and integrity in 23–36;
challenges in practice 120–128; delineating sensitive topics from similar
concepts 117–118; disciplinary debates
on research ethics in 6–7; gap of
research ethics between physical and
46–49; insights from two case studies
120–128; research ethics in 1–15,
42–55; research ethics in projects from
49–52; researching sensitive topics
118–120; sensitive research contexts
as multidisciplinary research field
115–117; sensitive topics in 114–130

identity 144–145, 156n6, 171–172
illegal ethnographies 153–164; 'after'
the field 160–162; epistemological
deadlock 153; in the field 158–160;
inception 156–158; overview 153–155

immanent critique 235–236
Indigenous knowledge 49–50, 55
individual values 204
inequalities of knowledge production:
　challenges of benefiting research
　participants 104–108; in Eastern
　Europe 92–109; in Eurasia 92–109;
　experiences in research field and in
　academia 96–100; navigating group
　discussions in Russian Karelia
　100–104; outsider in the native context
　104–108; power asymmetries 96–100
informed consent 66–70; and
　vulnerability 85–86
Institute of British Geographers 30
institutional codes of ethics 28
institutional ethics 5, 7, 30, 35, 106
institutionalised ethics 2–5, 47
institutionalised knowledge 203
institutional protocols 174
institutional requirements 1, 13–14
Institutional Review Board (IRB)
　161
integrity in human geography 23–36
interference 65, 68, 83–84, 117, 229,
　233–234
intersectionality 171
intervention 48, 51–52, 61, 69, 71, 95,
　103, 184
intuitively 217
Inuit community 49
Israel, M. 5, 7, 23–25, 28–29, 32–33,
　42

justice 156–157, 168; distributive 29;
　scarcity of 215; social 6, 24, 29;
　socio-spatial 135, 139–140

Karelia (Russia) 100–104
Kaufmann, F. 200, 203
Kindon, S. 7, 28
knowledge production: coloniality of
　93, 98, 106, 108; community-related
　9; conditions of 9–10; inequalities of
　92–109; uneven geographies of 93
Kyrgyzstan 104, 107

learning: ethical dimension of 213–221;
　explorative 213–214
limits of self-regulation 42–55
literacy, ethical 44, 54–55
local communities 34, 49–50, 121–122
locational privacy 28
long-term wellbeing 171

maltreatment 71–74
mapping 60, 186–190
mapping Muslim neighbourhoods
　186–190; analysis following the
　seven-step approach 187–190; check
　facts 187; develop a list of options
　188; identify relevant factors 188;
　review steps 1–6 190; select choice
　based on steps 1–5 190; state problem
　187; test options 188–189
marginalised communities 107
Markey, S. 160n9
marshrutka 106
memory 171
mental distress 25
mental health 27, 116, 177
mentoring 11
meta-ethics 7
México Indígena 25
Miggelbrink, J. 44n1
migration 9, 78–80, 215
misconduct: minimising and managing
　33; in research 34; scientific 31–32
moral anchors, responding to drifting
　34–35
moral authority 34–35
moral didactics 206–207
moral dilemma 200–211
moral economy of attention 5
moral geographies 6
morals 207
moral turn 6
moral values 24, 35, 207
Muslim communities 186–189

National Research Council of United
　States 4
neoliberal university 171–176; entering
　174–176; ethics and speed of research
　practice in 173–176
New Zealand Geographical Society 30
Nicaraguan youth gangs 119
non-contact sexual abuse 71
non-maleficence 140–142
normative ethics 7, 11, 207
normative reasoning 6
normative thinking 174
norms 62, 65, 70; ethical 14; socially con-
　structed management 51; Western 93
"not doing harm" 2

Office for Human Research Protections
　32
Olson, E. 6

ontological entanglements 229
organisational and professional
    demands 32–34
Orientalism 203

Pacific Island communities 31
Pain, R. 7, 136, 141
Palanpuri Jains 126
participant consent 47, 66–70
participation: children's rights to 59–61,
    65; voluntary 85, 139
participatory action research 6, 68, 134
patient privacy 28
patronage access 83
pedagogies 6, 70, 182, 184–185, 196–197
peer review 12n2, 32, 234–235
personal privacy 28
Pettig, F. 14, 214n2
phenomenological hermeneutics 215
phenomenology 230
physical anonymity 85
physical distress 115
physical geographers: and 'ethics
    of intervention' 48; research
    cooperation between human and 43;
    research ethics 44–46
physical geography: gap of research
    ethics between human and 46–49;
    research ethics in 42–55; research
    ethics in projects from 49–52
Pilz, M. 8, 96–100, 99n2
plagiarism 31
Poland 100, 121
politics of care 139–140, 145–147, 177
Popke, J. 6, 13, 15, 154, 177
position: ethical 67, 94, 153; of frontline
    practitioners 136; homogeneous 134;
    paradoxical 100; peripheral 173;
    political 97; shared conceptual 1;
    unethical 162
positionality 7, 64, 93, 101, 134, 159, 172,
    223, 226
postcolonial: approaches 202; studies
    for ethical guidance and orientation
    8
post-Enlightenment ethics 29
postmodernity 34
post-socialist: context 92, 100–101;
    Eastern Europe 92–94; Eurasia
    92, 94; literature 100; Poland 100;
    transformation in Georgia 98
post-Soviet decolonial thinking 93
post-war 169, 171; landscapes 172
power asymmetries 96–100

power in research processes 7–10
power relations 8, 49; asymmetric 84,
    95, 143; complex 141, 226; differential
    62; generational 66–67, 73; hierarchi-
    cal 224; political 96; problematic 72;
    social 69; unequal 81, 84, 87
practical ethics 2, 182–197
practitioners 29, 134–136, 138–145, 203
principle: of 'Do No Harm' 81, 85;
    ethical 3, 29, 42, 73, 138–139, 201–202,
    204, 206; of informed consent 3,
    13; of not causing harm 1; research
    ethics 61–66; in research ethics 2–4
principlism 29
privacy 201n1; changing perceptions
    of 193; locational 28; patient 28;
    personal 28
problem-centred interviews 123
procedural approach 2
procedural ethics 2, 12–13, 23, 137
Proctor, J. 6–7, 23, 47
profession 30, 133, 235
professionalisation 30
professional work 30–32; public trust
    and support for 30–32
profiling: racial 186–187; religious 186
protecting: others (and ourselves)
    24–27; researchers' rights 120;
    research subjects' rights 120
protests in Tbilisi 98–99
protocols: ethical 63, 137; institutional
    174; routine academic 135; security
    164
psychological distress 115
public communities 183
public trust 24, 30–32

qualitative methods 47, 102, 104;
    ethnographic 78
quantitative methods 3, 52,
    194

racial profiling 186–187
reciprocity 96–97
recognition 4, 30, 65–66, 82, 99, 169,
    171–172, 174, 176–177, 235
re-enacting and re-placing relationships
    145–147
reflection: as distancing 225–229; as
    distancing and turning 231–234
reflective practice 224
reflexive turn 81
reflexivity 11; communicative 87;
    ethical 65; feminist ethics of 232; and

feminist theory 225–226; retrospective 175; situated 64–66; transparent 86

refugee communities 83–84

refugees and asylum seekers: in Belarus and Ukraine 82–83

Rekhviashhvili, L. 104–108

relational ethics: geography and 137–138; situated reflexivity, principles and practices 64–66

relationships 133–147; re-enacting and re-placing 145–147; in research processes 7–10; thinking through 142–145

relativism 35

release of data 190–196; analysis following seven-step approach 190–192; designing project using automated digital diary 192–194; questionnaire to assess student attitudes 194–196

religious communities 125

religious profiling 186

reputation 1, 11, 12n2, 124–125

research: confidentiality 84; covert 29

researcher health 134, 172–174

researcher trauma 168–178, 171–176; defining 169–171; feminist praxis 171–173; feminist praxis and neoliberal university 171–176; overview 168–169; and research ethics 171–176; writing into research ethics and an ethics of care 176–177

research ethics 171–176; basic principles in 2–4; coping with challenging problems 27–29; and data handling 12–13; debates and education 29–30; disciplinary debates on, in human geography 6–7; in human geography 1–15, 23–36, 42–55; increasing sum of good 24–27; and inequalities of knowledge production in Eastern Europe 92–109; and inequalities of knowledge production in Eurasia 92–109; minimising harm 24–27; as part of research process 4–5; in physical geography 42–55; principles, practices and tensions 61–64; in projects from human and physical geography 49–52; to protect others (and ourselves) 24–27; public trust and support for professional work 30–32; respond to drifting moral anchors 34–35; satisfy organisational and professional demands 32–34;

and scientific education 14; and self-regulation 52–55; strengthening 52–55; in a wider neoliberal research context 10–12; writing researcher trauma into 176–177

research participants: challenges of benefiting 104–108

research principles 118, 201

research processes: power and relationships in 7–10; research ethics as part of 4–5

research quantum 175

resonance 217–218, 221

responsibility 2, 9; criminal 60; ethical 138, 162, 203, 229, 233; institutional 173; layers of 10; of peers 12n2; researcher's 85, 87, 104, 177; social 35

rights: children's voting 60; community-related 9; human 35; individual 9, 48, 118, 129; political and social 82; property 25; researchers' 106, 115, 120, 123–124, 127–128

risk 9–11, 83–84, 87, 176; behaviour 116; of personal liability 33; predictable 65; security-related risks of fieldwork 119; social and economic 125

roles of researchers 50, 55

Rose, G. 7, 154, 175, 224, 226–227, 229, 231, 238

routine academic protocols 135

Royal Geographical Society 30

Russia 101, 103, 121, 123

Said, Edward 203

Schäfer, S 15, 44n1, 68

science scandals 42

scientific communities 2, 42, 183

scientific education: and research ethics 14; strengthening ethical reasoning through 1

scientific knowledge 50, 203; acquisition 203; evaluating and transmitting 50; production 12, 47

scientific misconduct 31–32

secondary stress disorder 170

'secondary use' 12

Second World War 24

security protocols 164

self-care 175, 177

self-regulation: and external regulation 53; individual 52–53; limits of 42–55; of members' ethical conduct 30; need for 31; and strengthening research ethics 52–55 sensitive: research contexts as multi-disciplinary research field 115–117; topics in human geography 114–130

sensitive topics: appropriate methodological framework 118–120; challenges related to researching 118–120; delineating from similar concepts 117–118; in human geography 114–130; protecting research subjects' and researchers' rights 120

sensitivity 5; contextual 13; ethical 10; towards global dependencies 108; towards global hierarchies 108; universal 30

seven-step approach for teaching 184–185; examples 186–196

sexual abuse: contact 71; non-contact 71

short-term wellbeing 171

Silverman, D. 14

situated knowledge 133, 172, 174, 225

situated reflexivity 64–66

situations 2, 25–27, 62, 65, 67, 69, 73, 97–98, 153, 155, 159, 182, 184, 188, 201–202, 205–207, 211, 215, 217, 219

Skelton, T. 60, 63, 67, 74

slow scholarship 175

Smith, D. 6–7, 47

social justice 6, 24, 29

socially constructed management norms 51

socially marginalised communities 59

socially underprivileged communities 169

social values 24, 35

socio-spatial childhood research: ethics as a question of 'how' *and* 'what' in 59–75

socio-spatial justice 135, 139–140

space 1, 6–7, 78; of exception 236–237; illegal 157; shared symbolical 98; urban 28

spatial: injustices 7; justice 139; planning 206; representation 6; sciences 61, 92

'spatial turn' 231

stereotypes 70, 97, 100–101

subject matter 66, 214–220

survey 42–44, 44n1, 46, 48, 70, 126, 175, 182, 194–195

sustainability 50, 55, 215

Suyarkulova, M. 92, 95

symptoms 170–171

taboo 72, 114, 116, 129, 154, 163

tacit knowledge 217

teaching: ethical dimension of 213–221; ethical research principles to geography students 200–211

third convergence 43, 49

time 50; investing 98; and research ethics 108–109; stealing 108–109; and Western academia 98

Tlostanova, M. 92–94, 100

trader communities 126

training 50, 65, 72, 74, 93, 104, 136, 138–140, 147, 163, 169, 173, 177, 183, 216

transformation societies 94

trauma 81, 105; researcher 168–178; vicarious 170–171

trauma-informed research practice 177

trust: coalescence of 144; public 24, 30–32; unwarranted 9

turning: and attention 229–231; and ethics of academic reviewing 234–237; reflection as 231–234

Tuskegee syphilis study 24

Ukraine 82–83; practising research with refugees and asylum seekers in 82–83; qualitative fieldwork in 79

UK Research and Innovation (UKRI) 4

uneven geographies of knowledge production 93

United States: Belmont Report 29; Institutional Review Board (IRB) committees in 161; institutions 32; National Geographic Society 30; National Research Council of 4

universal sensitivity 30

Valentine, G. 7, 23, 154

values: ethical 139; individual 204; moral 24, 35, 207; social 24, 35

vicarious trauma 170–171

'vicarious traumatisation' 115

volunteering 147; praxis 134; roles 142; work 134

volunteer-practitioner research 133–147; doing research as volunteer-practitioners 139–142; ethics in helping and caring professions 138–139; friendship-liness 142–145; geographical research and helping and caring professions 135–136; geography and relational ethics

137–138; re-enacting and re-placing relationships 145–147; thinking through relationships 142–145

volunteer-practitioners: doing research as 139–142; non-maleficence and beneficence 140–142

vulnerability: role of gatekeepers and 83–85; role of informed consent and 85–86

vulnerable: to exploitation 79; groups 78–81; participants 79, 137; refugees and asylum seekers as 80–81

vulnerable groups: concepts and implications 79–81; researching 79–81

wellbeing 69; child 74; emotional 169; long-term 171; short-term 171

Western academic context 92, 100, 104, 108

Western academic ethical standards 93

Western academic knowledge production 96

Western ethical thinking 95

Western norms 93

Whitehead, A. N. 196

Whitehead, M. 146

Willowbrook State School hepatitis experiments 24

Wintzer, J. 6, 14, 214

work: academic 4; community 135; empirical 10, 61, 97, 123; ethnographic 154, 162; health and safety; human geographic 25, 27; illegal 127; interpersonal professional 138; professional 30–32; social 135; volunteering 134; youth 135, 144–146

workforce 169, 175

youth: clubs 68; cultures 119; Nicaraguan youth gangs 119; work 135, 144–146

Printed in the United States
by Baker & Taylor Publisher Services

Printed in the United States
by Baker & Taylor Publisher Services